Studies in Advanced Mathematics

An Introduction to Operator Algebras

Studies in Advanced Mathematics

Titles Included in the Series

Real Analysis and Foundations, *Steven G. Krantz*

CR Manifolds and the Tangential Cauchy–Riemann Complex, *Albert Boggess*

Elementary Introduction to the Theory of Pseudodifferential Operators, *Xavier Saint Raymond*

Fast Fourier Transforms, *James S. Walker*

Measure Theory and Fine Properties of Functions, *L. Craig Evans and Ronald Gariepy*

Partial Differential Equations and Complex Analysis, *Steven G. Krantz*

The Cauchy Transform, Potential Theory, and Conformal Mapping, *Steven R. Bell*

Several Complex Variables and the Geometry of Real Hypersurfaces, *John P. D'Angelo*

Modern Differential Geometry of Curves and Surfaces, *Alfred Gray*

An Introduction to Operator Algebras, *Kehe Zhu*

Wavelets: Mathematics and Applications, *John J. Benedetto and Michael W. Frazier*

KEHE ZHU
State University of New York at Albany

An Introduction to Operator Algebras

CRC PRESS
Boca Raton   Ann Arbor   London   Tokyo

Library of Congress Cataloging-in-Publication Data

Zhu, Kehe 1961–
An introduction to operator algebras / Kehe Zhu.
p. cm. — (Studies in advanced mathematics)
Includes bibliographical references and index.
ISBN 0-8493-7875-3
1. Operator algebras. I. Title. II. Series.
QA326.Z48 1993
512′.55—dc20 93-7172
CIP

Direct all inquiries to CRC Press, Inc., 2000 Corporate Blvd., N.W., Boca Raton, Florida 33431.

International Standard Book Number 0-8493-7875-3

Printed in the United States of America 1 2 3 4 5 6 7 8 9 0

Printed on acid-free paper

To Peijia and Peter

Contents

Preface

This is a graduate text for a one semester course in operator algebras. I taught such a course in the spring semester of 1992 at SUNY-Albany and this book represents a slightly expanded version of the lecture notes that I prepared during the semester. Since the theory of operator algebras has become such a vast subject, a difficult task the instructor faces when teaching a one semester course is deciding what topics to cover. Obviously the decision has to be somewhat personal. I hope my way of choosing the topics would be of some help for future instructors.

The book consists of twenty-six lectures. The two lectures on non-unital Banach and C^*-algebras can be omitted without affecting any other part of the book. Almost all the exercises are elementary, so that they are appropriate for homework assignments.

The latex style file for this book-series uses two different systems of numbering in each chapter. Theorems, propositions, lemmas, corollaries, and remarks are numbered as one group, while definitions and examples are numbered as another.

I would like to thank all the students who attended my Spring 1992 class. They are Robert Clark, Eric Grossman, Donghan Luo, Kevin Madigan, Jim Panariello, John Racquet, Jose Reyes, and John Rice. I am grateful to Steven Krantz and Norberto Salinas for their interest in the book project. Thanks are also due to Wayne Yuhasz, the CRC Press executive editor in Physical Sciences, for his enthusiasm and technical assistance in preparing the camera-ready manuscript. Finally I would like to thank the National Science Foundation for its support during the preparation of the manuscript.

Part I

BANACH ALGEBRAS

1

Review on Functional Analysis

1.1

Let $\mathbf{C}$ be the space of all complex numbers and let $\mathbf{R}$ be the space of all real numbers. $\mathbf{R}^+$ and $\mathbf{R}^-$ will denote the nonnegative and nonpositive numbers, respectively. All vector (or linear) spaces in the book will be over the field of complex numbers. In particular, all algebras will be over the complex numbers.

The reader is assumed to be familar with the basics of general topology, abstract algebra, real analysis, complex analysis, and functional analysis. The following is a review of several possibly less familar topics in functional analysis that will be used later in the book. All unproven results in this section can be found in any standard functional analysis text. We mention [9][5] in particular.

1.2 Riesz Representation

The phrase "Riesz representation" means one of at least three theorems in analysis. The following is a brief account of them.

THEOREM 1.1

F is a bounded linear functional on a Hilbert space H if and only if there exists a vector y in H such that $F(x) = \langle x, y\rangle$, $x \in H$, where $\langle\ ,\ \rangle$ is the inner product in H.

For a Hilbert space H we let $B(H)$ denote the space of all bounded linear operators on H. $B(H)$ is an algebra with the obvious linear structure and composition as multiplication. $B(H)$ is also a Banach space with the following operator norm

$$\|T\| = \sup\{\|Tx\| : \|x\| \leq 1\}, \qquad T \in B(H).$$

It follows easily from Riesz representation that for every T in $B(H)$ there exists a unique T^* in $B(H)$ such that

$$\langle Tx, y\rangle = \langle x, T^*y\rangle, \qquad x,\, y \in H.$$

The operator T^* is called the adjoint of T and the mapping $T \mapsto T^*$ has the following properties:

(a) $(aS + bT)^* = \bar{a}S^* + \bar{b}T^*$ for all $a, b \in \mathbf{C}$ and $S, T \in B(H)$.

(b) $(ST)^* = T^*S^*$ for all $S, T \in B(H)$.

(c) $(T^*)^* = T$ for all $T \in B(H)$.

(d) $\|T^*T\| = \|T\|^2$ for all $T \in B(H)$.

Suppose H is a Hilbert space and $F(x, y)$ is a function from $H \times H$ to $\mathbf{C}$. We say that F is a bounded bilinear form on H if it is linear in x, conjugate linear in y, and $|F(x, y)| \le C\|x\|\,\|y\|$ for some constant C and all x, y in H.

Suppose H is a Hilbert space and T is in $B(H)$. It is clear that the function $F: H \times H \to \mathbf{C}$ defined by $F(x, y) = \langle Tx, y\rangle$ is a bounded bilinear form on H. It turns out that all bounded bilinear forms on a Hilbert space arise in this fashion. In fact, if F is a bounded bilinear form on H, then for every fixed y in H the function $x \mapsto F(x, y)$ is a bounded linear functional on H, so that by the first version of Riesz representation there exists a vector in H, denoted Sy, such that $F(x, y) = \langle x, Sy\rangle$ for all x in H. It is easy to check that S defines a bounded linear operator on H. Letting $T = S^*$, we conclude that $F(x, y) = \langle Tx, y\rangle$ for all x and y in H. This proves the following result, also called the Riesz representation theorem.

THEOREM 1.2

A function $F : H \times H \to \mathbf{C}$ is a bounded bilinear form if and only if there exists a bounded linear operator T on H such that $F(x, y) = \langle Tx, y\rangle$ for all x and y in H.

Suppose K is a compact Hausdorff space. Let $\mathbf{C}(K)$ be the space of all complex-valued continuous functions on K. $\mathbf{C}(K)$ is a Banach space with pointwise operations and the sup-norm

$$\|f\| = \sup\big\{|f(x)| : x \in K\big\}, \qquad f \in \mathbf{C}(K).$$

The following is another useful result that bears the name Riesz representation.

THEOREM 1.3

Suppose K is a compact Hausdorff space. Then F is a bounded linear functional on $\mathbf{C}(K)$ if and only if there exists a finite regular complex Borel measure μ on

K such that

$$F(f)=\int_K f\,d\mu, \qquad f\in \mathbf{C}(K).$$

Suppose Ω is a locally compact Hausdorff space and $\mathbf{C}_0(\Omega)$ is the space of complex-valued continuous functions in Ω which can be uniformly approximated by continuous functions with compact support. Then every continuous linear functional F on $\mathbf{C}_0(\Omega)$ can be represented in the form

$$F(f)=\int_\Omega f\,d\mu, \qquad f\in \mathbf{C}_0(\Omega),$$

where μ is a finite regular complex Borel measure on Ω.

1.3 Weak Topologies

Suppose X is a set and $\mathcal{F}$ is a family of functions from X into a topological space Y. The weak topology on X induced by $\mathcal{F}$ is the weakest (smallest) topology on X which makes every function in $\mathcal{F}$ continuous. Thus a net $\{x_\alpha\}$ in X converges to some x in X in this weak topology if and only if $\{f(x_\alpha)\}$ converges to $f(x)$ for every f in $\mathcal{F}$.

Let X be a Banach space and let X^* be the Banach dual of X. By the weak topology on X we always mean the weak topology induced by the family of all bounded linear functionals on X. Thus a net $\{x_\alpha\}$ converges to x in X weakly (i.e., in the weak topology) if and only if $\{F(x_\alpha)\}$ converges to $F(x)$ for every F in X^*. The weak-star topology (or W^*-topology) on X^* is simply the weak topology on X^* induced by the family $\mathcal{F}=\{f_x : x\in X\}$, where for each x in X the function $f_x : X^*\to \mathbf{C}$ is defined by $f_x(F)=F(x)$, $F\in X^*$. Thus a net $\{F_\alpha\}$ in X^* converges to F in X^* in the weak-star topology if and only if $\{F_\alpha(x)\}$ converges to $F(x)$ for every x in X.

If X is a Banach space, it is easy to check that both the weak topology on X and the weak-star topology on X^* are Hausdorff.

THEOREM 1.4 ALAOGLU'S THEOREM
Suppose X is a Banach space and X^* is its dual space. Let $(X^*)_1$ be the closed unit ball in X^* (consisting of bounded linear functionals on X whose norms are less than or equal to 1). Then $(X^*)_1$ is compact in the weak-star topology.

PROOF We prove the compactness of $(X^*)_1$ in the weak-star topology by identifying it with a closed set in a compact space.

For each x in X let

$$\mathbf{D}_x=\{z\in\mathbf{C} : |z|\le\|x\|\}$$

with the usual Euclidean topology. Clearly each $\mathbf{D}_x$ is compact. By Tychonorff's theorem,

$$P = \prod_{x \in X} \mathbf{D}_x$$

is compact in the product topology. Define a mapping $\xi : (X^*)_1 \to P$ by

$$\xi(F)(x) = F(x), \qquad F \in (X^*)_1.$$

It is clear that ξ is one-to-one. Also, $F_\alpha \to F$ in $(X^*)_1$ in the weak-star topology if and only if $F_\alpha(x) \to F(x)$ for every x in X if and only if $\xi(F_\alpha) \to \xi(F)$ in P in the product topology. Thus ξ is a homeomorphism from $(X^*)_1$ onto its image in P. It remains to show that this image is closed in P.

Suppose $\xi(F_\alpha) \to p$ in P. Then $\lim_\alpha F_\alpha(x) = p(x)$ for every x in X. It is easy to check that $F : X \to \mathbf{C}$ defined by $F(x) = p(x)$ is a bounded linear functional with norm less than or equal to one. Thus we have $p = \xi(F)$ and hence the image of $(X^*)_1$ in P under the mapping ξ is closed in the product topology. ∎

1.4 Topological Vector Spaces

A space X is called a topological vector space if X is a vector space together with a Hausdorff topology such that vector addition and scalar multiplication are both continuous.

A topological vector space is called locally convex if the origin of X has a local base whose elements are convex.

Suppose X is a vector space and $\{\| \ \|_\alpha\}$ is a separating family of semi-norms on X. ("Separating" means that for each nonzero vector x in X there exists some α such that $\|x\|_\alpha \neq 0$.) The topology on X generated by $\{\| \ \|_\alpha\}$ is the unique Hausdorff topology on X such that $x_\beta \to x$ in X if and only if $\|x_\beta - x\|_\alpha \to 0$ for every α. The following sets constitute a local base for the origin in X:

$$\left\{x \in X : \|x\|_{\alpha_1} < \epsilon_1, \cdots, \|x\|_{\alpha_n} < \epsilon_n\right\}, \qquad \epsilon_k > 0,\ 1 \le k \le n,\ n \ge 1.$$

It is clear that X is locally convex with the topology generated by $\{\| \ \|_\alpha\}$. The following theorem shows that all locally convex spaces arise in this fashion.

THEOREM 1.5

A topological vector space X is locally convex if and only if its topology is generated by a separating family of semi-norms.

Note that if X is a locally convex topological vector space with its topology generated by a separating family of semi-norms $\{\| \ \|_\alpha\}$, then a linear functional

F on X is continuous if and only if $F(x_\beta) \to 0$ whenever $\|x_\beta\|_\alpha \to 0$ for every α.

THEOREM 1.6

Suppose X is a locally convex topological vector space with its topology generated by a separating family of semi-norms $\{\| \ \|_\alpha\}$. Then a linear functional F on X is continuous if and only if there exists a constant $C > 0$ and $\alpha_1, \cdots, \alpha_n$ such that

$$|F(x)| \leq C \max\{\|x\|_{\alpha_1}, \cdots, \|x\|_{\alpha_n}\}$$

for all x in X.

THEOREM 1.7

Suppose X is a locally convex topological vector space and S is a convex set in S. An element x in X belongs to the closure of S if and only if there exists a net $\{x_\alpha\}$ in S such that $F(x_\alpha) \to F(x)$ for every continuous linear functional F on X.

1.5 Extreme Points

Suppose X is a locally convex topological vector space and S is a convex subset of X. A point x in S is called an extreme point of S if x cannot be written as $x = tx_1 + (1-t)x_2$, with $t \in (0,1)$ and x_1, x_2 being different points in S. The following theorem assures the existence of extreme points for compact convex subsets of a locally convex space.

THEOREM 1.8 KREIN-MILMAN'S THEOREM

Let S be a compact convex set in a locally convex space X. Then the set E of extreme points of S is nonempty. Furthermore, S is the closed convex hull of E, namely, S is the closure of the set of all finite combinations $a_1x_1 + \cdots + a_nx_n$, where $x_1, \cdots, x_n$ are from E and $a_1, \cdots, a_n$ are positive numbers with $a_1 + \cdots + a_n = 1$.

1.6 The Stone-Weierstrass Approximation Theorem

THEOREM 1.9

Suppose K is a compact Hausdorff space. If $\mathcal{A}$ is a closed self-adjoint subalgebra of $\mathbf{C}(K)$ which contains the constant functions and separates the points of K, then $\mathcal{A} = \mathbf{C}(K)$.

1.7 Zorn's Lemma

Let X be a set. A partial ordering of X is a relation $\leq$ among some pairs of elements of X with the following properties:

(a) $x \leq x$ for all $x \in X$.

(b) If $x \leq y$ and $y \leq z$, then $x \leq z$.

(c) If $x \leq y$ and $y \leq x$, then $x = y$.

If for every pair x, y in X we have $x \leq y$ or $y \leq x$, we say that X is totally ordered and $\leq$ is a total ordering of X.

Suppose X is partially ordered by $\leq$ and S is a subset of X. We say that x is an upper bound of S if $y \leq x$ for all y in S. An element x in X is maximal if $x \leq y$ implies $x = y$. Note that maximal elements are not unique in general.

THEOREM 1.10 ZORN'S LEMMA
Suppose X is a nonempty set partially ordered by $\leq$. If every nonempty totally ordered subset of X has an upper bound, then X has a maximal element.

Exercises

1.1 Refresh your memory of real analysis and general topology. Pay special attention to regular Borel measures, compact Hausdorff spaces, Uryson's lemma, Tychonorff's theorem, and connectedness of topological spaces.

2

Banach Algebras and the Invertible Group

2.1 Banach Algebras

In this lecture we introduce the notion of a Banach algebra and study the (multiplicative) group of invertible elements. Again all algebras will be over the field of complex numbers. We also assume that all algebras are associative. An algebra is called unital if it has a (multiplicative) unit. Note that the unit is unique (if exists). We shall use $\mathbf{1}$ to denote the unit in a unital algebra.

DEFINITION 2.1 *A Banach algebra is a unital algebra $\mathcal{A}$ together with a complete norm $\| \ \|$ satisfying the following conditions:*

(a) $\|\mathbf{1}\| = 1$.

(b) $\|xy\| \leq \|x\| \, \|y\|$ *for all x and y in $\mathcal{A}$.*

2.2 Examples

The simplest Banach algebra is the complex field $\mathbf{C}$ with $\|z\| = |z|$ (the modulus of z). The following are several other examples of Banach algebras.

EXAMPLE 2.2

If K is a compact Hausdorff space, then $\mathbf{C}(K)$ is a Banach algebra with the sup-norm and pointwise operations. The constant function with value 1 is the multiplicative unit. $\square$

EXAMPLE 2.3

If H is a Hilbert space, then $B(H)$ is a Banach algebra with the operator norm and operator multiplication (i.e., composition). The identity operator I is the multiplicative unit.

Note that when $H = \mathbf{C}^n$ is finite dimensional, $B(\mathbf{C}^n)$ can be identified with the algebra $M_n(\mathbf{C})$ of all $n \times n$ matrices with complex entries. Under this identification, operator multiplication corresponds to matrix multiplication. $\Box$

EXAMPLE 2.4

Let $\mathbf{D}$ be the open unit disk in the complex plane $\mathbf{C}$. Then $H^\infty(\mathbf{D})$, the space of all bounded analytic functions in $\mathbf{D}$, is a Banach algebra with the sup-norm and pointwise operations. $\Box$

EXAMPLE 2.5

Let $\mathbf{Z}$ be the set of all integers with counting measure. Then $L^1(\mathbf{Z})$ is a Banach algebra with multiplication defined as follows:

$$f * g(n) = \sum_{k=-\infty}^{\infty} f(n-k)g(k), \qquad n \in \mathbf{Z},\ f,\ g \in L^1(\mathbf{Z}).$$

By Fubini's theorem,

$$\begin{aligned} \|f * g\| &= \sum_{n=-\infty}^{\infty} \left| \sum_{k=-\infty}^{\infty} f(n-k)g(k) \right| \\ &\le \sum_{n=-\infty}^{\infty} \sum_{k=-\infty}^{\infty} |f(n-k)||g(k)| \\ &= \sum_{k=-\infty}^{\infty} |g(k)| \sum_{n=-\infty}^{\infty} |f(n-k)| \\ &= \|f\|\,\|g\|. \end{aligned}$$

Thus the multiplication is well defined. It is easy to check that the above multiplication makes $L^1(\mathbf{Z})$ a unital commutative algebra. The multiplicative unit of $L^1(\mathbf{Z})$ is the function $\mathbf{1}$ defined by $\mathbf{1}(0) = 1$, $\mathbf{1}(n) = 0$ for $n \in \mathbf{Z} - \{0\}$. $\Box$

EXAMPLE 2.6

If $\mathcal{A}$ is a Banach algebra and $\mathcal{I}$ is a closed two-sided ideal of $\mathcal{A}$, then the quotient algebra $\mathcal{A}/\mathcal{I}$ is a Banach algebra with the following (quotient) norm:

$$\big\|[x]\big\| = \inf\big\{\|x-y\| : y \in \mathcal{I}\big\}, \qquad [x] \in \mathcal{A}/\mathcal{I}.$$

Since the set $\mathcal{K}$ of all compact operators on a Hilbert space H is a closed two-sided ideal of $B(H)$, the quotient algebra $Q(H) = B(H)/\mathcal{K}$ is a Banach algebra. $Q(H)$ is called the Calkin algebra on H. □

2.3 The Invertible Group

DEFINITION 2.7 *For a Banach algebra $\mathcal{A}$ we let $G(\mathcal{A})$ denote the set of all invertible elements in $\mathcal{A}$. It is clear that $G(\mathcal{A})$ is a multiplicative group.*

If K is a compact Hausdorff space, the invertible group of $\mathbf{C}(K)$ consists exactly of those nonvanishing functions. A function $f \in H^\infty(\mathbf{D})$ is invertible if and only if there exists a positive constant ϵ such that $|f(z)| \geq \epsilon$ for all z in $\mathbf{D}$. Suppose T is in $B(H)$ for some Hilbert space H. Then T is invertible in $B(H)$ if and only if T is one-to-one and onto; the inverse is guarenteed to be bounded by the open mapping theorem. When H is finite dimensional, $T \in B(H)$ is invertible if and only if T is one-to-one, a well-known fact in linear algebra.

For the remaining two examples in **2.2**, describing the invertible elements is a more interesting problem. In the case of the convolution algebra $L^1(\mathbf{Z})$, a two-sided sequence $f = \{f(n) : n \in \mathbf{Z}\}$ of complex numbers is invertible in $L^1(\mathbf{Z})$ if and only if the continuous function $\sum_{n\in\mathbf{Z}} f(n)z^n$ on the unit circle is nonvanishing. See Corollary 5.2 and Theorem 6.3. For the Calkin algebra $Q(H)$, invertible elements are precisely those cosets induced by Fredholm operators (namely, bounded linear operators on H with closed range and finite dimensional kernel and cokernel). This is called Atkinson's theorem in operator theory; see [2].

PROPOSITION 2.1

If $\mathcal{A}$ is a Banach algebra and x is in $\mathcal{A}$ with $\|x\| < 1$, then $\mathbf{1} - x$ is in $G(\mathcal{A})$ with $\|(\mathbf{1} - x)^{-1}\| \leq (1 - \|x\|)^{-1}$.

PROOF Let

$$y = \sum_{n=0}^{\infty} x^n = \lim_{N\to\infty} \sum_{n=0}^{N} x^n,$$

where $x^0 = \mathbf{1}$. The series (and limit) converges in $\mathcal{A}$ because

$$\left\| \sum_{n=0}^{N+p} x^n - \sum_{n=0}^{N} x^n \right\| \leq \sum_{n=N+1}^{N+p} \|x\|^n \to 0, \quad (N \to \infty)$$

uniformly in $p\,(\geq 1)$. Since mulitiplication is continuous in a Banach algebra, we have

$$(\mathbf{1}-x)y = y - xy = \sum_{n=0}^{\infty} x^n - \sum_{n=0}^{\infty} x^{n+1} = \mathbf{1}$$

and

$$y(\mathbf{1}-x) = y - yx = \sum_{n=0}^{\infty} x^n - \sum_{n=0}^{\infty} x^{n+1} = \mathbf{1}.$$

Thus $(\mathbf{1}-x)^{-1} = y$ and

$$\|(\mathbf{1}-x)^{-1}\| \leq \sum_{n=0}^{\infty} \|x\|^n = \frac{1}{1-\|x\|}$$

whenever $\|x\| < 1$. ∎

PROPOSITION 2.2

For each Banach algebra $\mathcal{A}$ the invertible group $G(\mathcal{A})$ is an open set in $\mathcal{A}$.

PROOF Suppose x is in $G(\mathcal{A})$. We show that the set

$$\left\{ y \in \mathcal{A} : \|y - x\| < \frac{1}{\|x^{-1}\|} \right\}$$

is contained in $G(\mathcal{A})$; this will clearly prove that $G(\mathcal{A})$ is open. Given y in $\mathcal{A}$ with $\|y-x\| < \|x^{-1}\|^{-1}$, it is clear that $\|x^{-1}\|\,\|y-x\| < 1$, and hence $\|x^{-1}(y-x)\| < 1$ or $\|\mathbf{1} - x^{-1}y\| < 1$. By Proposition 2.1, $x^{-1}y$ is invertible, so y is invertible. ∎

PROPOSITION 2.3

Let $\mathcal{A}$ be a Banach algebra. The inversion $x \mapsto x^{-1}$ is continuous on $G(\mathcal{A})$.

PROOF Suppose $x_n \to x$ in $G(\mathcal{A})$. Without loss of generality we may assume that for every n

$$\|x_n - x\| < \frac{1}{2\|x^{-1}\|},$$

so that

$$\|\mathbf{1} - x^{-1}x_n\| < \frac{1}{2}.$$

Since

$$\sum_{k=0}^{\infty} (\mathbf{1} - x^{-1}x_n)^k = x_n^{-1}x,$$

we have

$$\|x_n^{-1}x\| \leq \sum_{k=0}^{\infty} \frac{1}{2^k} = 2.$$

It follows that

$$\|x_n^{-1}\| \leq \|x_n^{-1}x\| \, \|x^{-1}\| \leq 2\|x^{-1}\|,$$

and

$$\|x_n^{-1} - x^{-1}\| = \|x_n^{-1}(x_n - x)x^{-1}\| \leq 2\|x^{-1}\|^2\|x_n - x\|,$$

finishing the proof of the proposition. ∎

2.4 The Abstract Index

THEOREM 2.4

Let $\mathcal{A}$ be a Banach algebra and let $G_0(\mathcal{A})$ be the connected component of $G(\mathcal{A})$ containing **1**. *Then $G_0(\mathcal{A})$ is an open, closed, and normal subgroup of $G(\mathcal{A})$. The cosets of $G_0(\mathcal{A})$ are exactly the connected components of $G(\mathcal{A})$. The quotient group $G(\mathcal{A})/G_0(\mathcal{A})$ is a discrete group.*

PROOF Since $G(\mathcal{A})$ is an open set in the connected space $\mathcal{A}$, it must be locally connected. Each connected component of a locally connected space is both open and closed. In particular, $G_0(\mathcal{A})$ is both open and closed in $G(\mathcal{A})$.

To show that $G_0(\mathcal{A})$ is a subgroup of $G(\mathcal{A})$, let x and y be elements in $G_0(\mathcal{A})$. The set $G_0(\mathcal{A}) \cup xG_0(\mathcal{A})$, being the union of two connected open sets with a nontrivial intersection, must be connected. It follows that $xG_0(\mathcal{A}) \subset G_0(\mathcal{A})$ and so $xy \in xG_0(\mathcal{A}) \subset G_0(\mathcal{A})$. Similary, $x^{-1}G_0(\mathcal{A}) \cup G_0(\mathcal{A})$ is the union of two connected open sets with a nontrivial intersection and must be connected. This implies $x^{-1}G_0(\mathcal{A}) \subset G_0(\mathcal{A})$ and $x^{-1} \in x^{-1}G_0(\mathcal{A}) \subset G_0(\mathcal{A})$. Thus $G_0(\mathcal{A})$ is closed under multiplication and inversion, namely, $G_0(\mathcal{A})$ is a subgroup of $G(\mathcal{A})$.

If x is in $G(\mathcal{A})$, $xG_0(\mathcal{A})x^{-1}$ is a connected open set containing **1**. This implies that $xG_0(\mathcal{A})x^{-1} \subset G_0(\mathcal{A})$. Applying the mapping $y \mapsto x^{-1}yx$ to the inclusion above, we conclude that $G_0(\mathcal{A}) \subset x^{-1}G_0(\mathcal{A})x$. Replacing x by x^{-1}, we obtain $G_0(\mathcal{A}) \subset xG_0(\mathcal{A})x^{-1}$ and hence $G_0(\mathcal{A}) = xG_0(\mathcal{A})x^{-1}$. Thus $G_0(\mathcal{A})$ is a normal subgroup of $G_0(\mathcal{A})$.

Finally, since each $xG_0(\mathcal{A})$ is connected, closed, and open in $G(\mathcal{A})$, the cosets of $G_0(\mathcal{A})$ in $G(\mathcal{A})$ are precisely the components of $G(\mathcal{A})$. It follows that $G(\mathcal{A})/G_0(\mathcal{A})$ is a discrete group. ∎

DEFINITION 2.8 *For a Banach algebra $\mathcal{A}$ the group $I(\mathcal{A}) = G(\mathcal{A})/G_0(\mathcal{A})$ is called the index group of $\mathcal{A}$ and the quotient mapping $\pi : G(\mathcal{A}) \to I(\mathcal{A})$ is called the abstract index of $\mathcal{A}$.*

The abstract index is an invariant of Banach algebras. Suppose Φ is a homomorphism from a Banach algebra $\mathcal{A}$ into another Banach algebra $\mathcal{B}$. It is easy to see that Φ induces a group homomorphism from $I(\mathcal{A})$ into $I(\mathcal{B})$. If $\mathcal{A}$ is isomorphic to $\mathcal{B}$, then $I(\mathcal{A})$ is isomorphic to $I(\mathcal{B})$.

2.5 Examples

The following are several examples of the abstract index.

EXAMPLE 2.9
Let $\mathcal{A} = \mathbf{C}(T)$, where T is the unit circle in the complex plane. It is clear that $G(\mathcal{A})$ consists of all nonvanishing functions in $\mathcal{A}$ and $G_0(\mathcal{A})$ consists of functions in $\mathcal{A}$ whose logarithms are also in $\mathcal{A}$. The index group of $\mathcal{A}$ is isomorphic to the group of integers, and the abstract index of a nonvanishing function in $\mathcal{A}$ is simply its topological winding number. $\square$

EXAMPLE 2.10
Let $\mathcal{A} = B(H)$ with H a Hilbert space. It is a nontrivial fact that $G_0(\mathcal{A}) = G(\mathcal{A})$ in this case (see Corollary 23.5). Thus the abstract index group of $B(H)$ is trivial. Note that this can be shown using elementary linear algebra when H is finite dimensional; see Exercise 2.2. $\square$

EXAMPLE 2.11
Let $\mathcal{A} = Q(H)$, the Calkin algebra of an infinite dimensional Hilbert space H. Recall that a bounded linear operator T on H is called Fredholm if T has closed range and both $\ker T$ and $\ker T^*$ are finite dimensional. The invertible group of $Q(H)$ consists of the cosets $[T]$, where T is Fredholm on H. The abstract index group of $Q(\mathcal{A})$ is again isomorphic to the integer group. Given $[T]$ in $G(\mathcal{A})$, the abstract index of $[T]$ coincides with the Fredholm index of T which is given by $\dim(\ker T^*) - \dim(\ker T)$. See [2]. $\square$

EXAMPLE 2.12
Let $\mathcal{A} = H^\infty(\mathbf{D})$, the algebra of all bounded analytic functions on the open unit disk $\mathbf{D}$. Then $G(\mathcal{A})$ coincides with $G_0(\mathcal{A})$. This follows from the easy fact that every nonzero analytic function in $\mathbf{D}$ has an analytic logarithm in $\mathbf{D}$. $\square$

EXAMPLE 2.13

Let $\mathcal{A} = L^1(\mathbf{Z})$, the convolution algebra over $\mathbf{Z}$. Using some results in Lectures 5 and 6, we can show that the abstract index group of $\mathcal{A}$ can again be identified with the integer group $\mathbf{Z}$. We omit the details. $\square$

Exercises

2.1 Verify the assertions made in Examples 2.9 and 2.12.

2.2 Verify the assertions made in Example 2.10 in the finite-dimensional case.

2.3 Let

$$\mathcal{A} = \{f \in \mathbf{C}[0,1] : f' \in \mathbf{C}[0,1]\}$$

with pointwise operations and

$$\|f\| = \|f\|_\infty + \|f'\|_\infty.$$

Show that $\mathcal{A}$ is a Banach algebra.

2.4 Suppose $\mathcal{A} = M_n(\mathbf{C})$. Show that

$$\begin{aligned} \|A\|^2 &= \max\{\lambda : \det(\lambda I - A^*A) = 0\} \\ &= \max\{\lambda : \det(\lambda - AA^*) = 0\}, \end{aligned}$$

where A^* is the conjugate transpose of A.

2.5 An operator in $B(\mathbf{C}^n)$ $(n < +\infty)$ is invertible if and only if it is one-to-one if and only if it is onto. Show that this is false for $B(H)$ with $\dim H = +\infty$.

2.6 Let X be a Banach space and $B(X)$ be the space of all bounded linear operators on X. Clearly $B(X)$ is a vector space. Show that $B(X)$ is a Banach algebra if the norm is defined by

$$\|T\| = \sup\{\|Tx\| : \|x\| \le 1\}$$

and if the multiplication is defined by composition.

2.7 Suppose $\mathcal{A}$ is a Banach algebra and $\mathcal{I}$ is a closed two-sided ideal in $\mathcal{A}$. Show that $\mathcal{A}/\mathcal{I}$ is a Banach algebra with the quotient norm.

3

The Spectrum

3.1 The Definition

In this lecture we introduce the notion of the spectrum of an element in a Banach alagebra and study the elementary properties of the spectrum. We can think of the spectrum as a generalization of the range of a function in the case of a commutative Banach algebra. In the general case the spectrum can be thought of as a generalization of the notion of eigenvalues.

DEFINITION 3.1 *Suppose $\mathcal{A}$ is a Banach algebra and x is in $\mathcal{A}$. The spectrum of x, denoted by $\sigma_{\mathcal{A}}(x)$ or simply by $\sigma(x)$, is the set of all complex numbers λ such that $\lambda\mathbf{1} - x$ is not invertible in $\mathcal{A}$. The compliment of $\sigma(x)$ in $\mathbf{C}$ is called the resolvant set of x.*

Note that the notion of the spectrum is purely algebraic; it only depends on the algebraic structure of the Banach algebra. However, what is essential for the usefulness of the spectral theory is the presence of a "compatible" Banach space structure, without which the spectrum of an element may not even exist.

3.2 Examples

EXAMPLE 3.2

Let $\mathcal{A} = \mathbf{C}(K)$ with K compact and Hausdorff. Then $\sigma(f) = f(K)$ for every f in $\mathcal{A}$. ▯

EXAMPLE 3.3

Let $\mathcal{A} = H^\infty(\mathbf{D})$. Then for each f in $\mathcal{A}$ the spectrum $\sigma(f)$ is the Euclidean closure of the image of f. □

EXAMPLE 3.4

Let $\mathcal{A} = B(\mathbf{C}^n) = M_n(\mathbf{C})$, the $n \times n$ matrix algebra. Then for each T in $\mathcal{A}$ the spectrum $\sigma(T)$ consists of all eigenvalues of the matrix T. This follows easily from linear algebra.

Let $\mathcal{A} = B(H)$, where H is an infinite dimensional Hilbert space, and let $T \in \mathcal{A}$. It is easy to see that $\sigma(T)$ contains all eigenvalues of the operator T. However, unlike the finite dimensional case, an operator in $B(H)$ does not necessarily have an eigenvalue. For example, if $H = L^2[0,1]$ with Lebesgue measure and if T is the multiplication operator induced by the identity function, namely, $Tf(x) = xf(x)$, then T belongs to $B(H)$ but T does not have an eigenvalue. Nevertheless, we shall see a little later that the spectrum $\sigma(T)$ is always nonempty. □

EXAMPLE 3.5

Let $\mathcal{A} = L^1(\mathbf{Z})$ be the convolution algebra and let f be in $\mathcal{A}$. We shall see later (Corollary 5.2 and Theorem 6.3) that

$$\sigma(f) = \left\{ \sum_{n \in \mathbf{Z}} f(n) e^{int} : t \in \mathbf{R} \right\}.$$

It is interesting to try to verify this directly. □

EXAMPLE 3.6

Let $\mathcal{A} = Q(H)$ be the Calkin algebra of a Hilbert space H and let T be in $B(H)$. The spectrum of $[T]$ in $\mathcal{A}$ is called the essential spectrum of the operator T. Thus a complex number λ belongs to the essential spectrum of T if and only if $\lambda I - T$ is not Fredholm. □

EXAMPLE 3.7

Suppose Φ is a homomorphism from a Banach algebra $\mathcal{A}$ into another Banach algebra $\mathcal{B}$ and suppose Φ maps the unit of $\mathcal{A}$ to the unit of $\mathcal{B}$. Then it is easy to show that $\sigma_{\mathcal{B}}(\Phi(x)) \subset \sigma_{\mathcal{A}}(x)$ for all x in $\mathcal{A}$. If Φ happens to be a surjective isomorphism, then $\sigma_{\mathcal{B}}(\Phi(x)) = \sigma_{\mathcal{A}}(x)$ for all x in $\mathcal{A}$. □

3.3 Basic Properties of the Spectrum

We first prove that the spectrum of an element in a Banach algebra is nonempty. This is one of the main reasons why the notion of spectrum is so useful in the theory of Banach algebras.

THEOREM 3.1
Suppose $\mathcal{A}$ is a Banach algebra and x is in $\mathcal{A}$. Then $\sigma(x)$ is nonempty.

PROOF If x is not invertible, its spectrum contains 0. Thus we may as well assume that x is invertible. Attempting to arrive at a contradiction, we assume that $\sigma(x)$ is empty. Fix a bounded linear functional F on $\mathcal{A}$ and consider the function $f : \mathbf{C} \to \mathbf{C}$ defined by

$$f(z) = F\big((z\mathbf{1} - x)^{-1}\big), \qquad z \in \mathbf{C}.$$

The function f is defined on the entire complex plane since $\sigma(x)$ is empty. Fix $z_0 \in \mathbf{C}$. By the continuity of F we have

$$\lim_{z \to z_0} \frac{f(z) - f(z_0)}{z - z_0} = F\left(\lim_{z \to z_0} \frac{(z\mathbf{1} - x)^{-1} - (z_0\mathbf{1} - x)^{-1}}{z - z_0}\right).$$

Since

$$(z\mathbf{1} - x)^{-1} - (z_0\mathbf{1} - x)^{-1} = (z\mathbf{1} - x)^{-1}(z_0 - z)(z_0\mathbf{1} - x)^{-1},$$

and the inversion is continuous, we obtain

$$\lim_{z \to z_0} \frac{(z\mathbf{1} - x)^{-1} - (z_0\mathbf{1} - x)^{-1}}{z - z_0} = -(z_0\mathbf{1} - x)^{-2},$$

and hence

$$\lim_{z \to z_0} \frac{f(z) - f(z_0)}{z - z_0} = -F\big((z_0\mathbf{1} - x)^{-2}\big).$$

We see that f is an entire function. Furthermore, if $|z| > \|x\|$, by Proposition 2.1

$$\begin{aligned} |f(z)| &\le \|F\| \, \|(z\mathbf{1} - x)^{-1}\| = \frac{\|F\|}{|z|} \|(\mathbf{1} - \frac{x}{z})^{-1}\| \\ &\le \frac{\|F\|}{|z|} \frac{1}{1 - \frac{\|x\|}{|z|}} = \frac{\|F\|}{|z| - \|x\|}. \end{aligned}$$

Thus $f(z)$ is a bounded entire function which converges to 0 as $|z| \to +\infty$. By Liouville's theorem, f must be identically zero. Since F is arbitrary, the Hahn-Banach extension theorem implies that $(z\mathbf{1} - x)^{-1} = 0$, which is impossible because 0 is never invertible in a Banach algebra. ∎

Recall that an algebra whose nonzero elements are invertible is called a division algebra. The following theorem is usually called the Gelfand-Mazur theorem.

THEOREM 3.2
If $\mathcal{A}$ is a Banach algebra in which every nonzero element is invertible, then $\mathcal{A}$ is isometrically isomorphic to $\mathbf{C}$.

PROOF Fix x in $\mathcal{A}$ and let $\lambda \in \sigma(x)$. Since $\lambda\mathbf{1} - x$ is not invertible, it must be 0, or $x = \lambda\mathbf{1}$. Define $\Phi : \mathbf{C} \to \mathcal{A}$ by $\Phi(z) = z\mathbf{1}$. Then Φ is linear, one-to-one, and onto. Furthermore, Φ preserves multiplication and norm. Thus Φ is an isometrical isomorphism. ∎

THEOREM 3.3
Suppose $\mathcal{A}$ is a Banach algebra and x is in $\mathcal{A}$. Then $\sigma(x)$ is compact in $\mathbf{C}$ *and is contained in the closed disk* $\{z \in \mathbf{C} : |z| \leq \|x\|\}$.

PROOF Fix x in $\mathcal{A}$. The complement of $\sigma(x)$ in $\mathbf{C}$, $\mathbf{C} - \sigma(x)$, is open, because $G(\mathcal{A})$ is open and the inversion is continuous. Thus $\sigma(x)$ is closed in $\mathbf{C}$.

For $|\lambda| > \|x\|$, the element

$$\lambda\mathbf{1} - x = \lambda(\mathbf{1} - \frac{x}{\lambda})$$

is invertible by Proposition 2.1. So $\sigma(x)$ is contained in the closed disk $\{z \in \mathbf{C} : |z| \leq \|x\|\}$. Being bounded and closed, $\sigma(x)$ is compact. ∎

3.4 The Spectral Radius

DEFINITION 3.8 *Suppose $\mathcal{A}$ is a Banach algebra and x is in $\mathcal{A}$. The number*

$$r(x) = \sup\{|\lambda| : \lambda \in \sigma(x)\}$$

is called the spectral radius of x.

By Theorem 3.3, we always have $0 \leq r(x) \leq \|x\|$. These estimates are best possible, and both extremes can occur. In fact, it is easy to see that $r(x) = \|x\|$ for all x if $\mathcal{A} = \mathbf{C}(K)$. On the other hand, if $\mathcal{A} = M_2(\mathbf{C})$, the 2×2 matrix algebra, and if

$$x = \begin{pmatrix} 0 & 1 \\ 0 & 0 \end{pmatrix},$$

then $\sigma(x) = \{0\}$ and hence $r(x) = 0$. In general, if x is in $\mathcal{A} = M_n(\mathbf{C})$, the $n \times n$ matrix algebra, then the spectral radius of x is the maximal modulus of the eigenvalues of x.

Suppose $\mathcal{A} = L^1(\mathbf{Z})$. Recall from Example 3.5 that for $f \in L^1(\mathbf{Z})$ the spectrum $\sigma(f)$ is the range of the function

$$\sum_{n\in\mathbf{Z}} f(n)e^{int}, \qquad t \in \mathbf{R}.$$

It follows that

$$r(f) = \sup\left\{\left|\sum_{n\in\mathbf{Z}} f(n)e^{int}\right| : t \in \mathbf{R}\right\}.$$

PROPOSITION 3.4

If x and y are in a Banach algebra $\mathcal{A}$, then $\mathbf{1} - xy$ is invertible if and only if $\mathbf{1} - yx$ is invertible.

PROOF If $\mathbf{1} - xy$ is invertible in $\mathcal{A}$, then the element

$$y(\mathbf{1} - xy)^{-1}x + \mathbf{1}$$

is the inverse of $\mathbf{1} - yx$. In fact,

$$\begin{aligned}
&(\mathbf{1} - yx)\big[y(\mathbf{1} - xy)^{-1}x + \mathbf{1}\big] \\
&\quad = y(\mathbf{1} - xy)^{-1}x + \mathbf{1} - yxy(\mathbf{1} - xy)^{-1}x - yx \\
&\quad = y\big[(\mathbf{1} - xy)^{-1} - xy(\mathbf{1} - xy)^{-1}\big]x + \mathbf{1} - yx \\
&\quad = yx + \mathbf{1} - yx = \mathbf{1}.
\end{aligned}$$

Similarly,

$$\big[y(\mathbf{1} - xy)^{-1}x + \mathbf{1}\big](\mathbf{1} - yx) = \mathbf{1}.$$

This completes the proof of the proposition. ∎

COROLLARY 3.5

Suppose $\mathcal{A}$ is a Banach algebra and $x, y \in \mathcal{A}$. Then

$$\sigma(xy) \cup \{0\} = \sigma(yx) \cup \{0\}.$$

COROLLARY 3.6

Suppose $\mathcal{A}$ is a Banach algebra and $x, y \in \mathcal{A}$. Then $r(xy) = r(yx)$.

Exercises

3.1 Let T be the operator defined on $H = L^2[0,1]$ by

$$Tf(x) = xf(x), \qquad x \in [0,1].$$

What is the spectrum of T in $B(H)$? Generalize your result.

3.2 Suppose H is a (separable) infinite dimensional Hilbert space. Show that for every nonempty compact set K in the complex plane there exists an operator T in $B(H)$ such that $\sigma(T) = K$.

3.3 Let T be the operator on l^2 defined by

$$T(\{a_1, a_2, \cdots\}) = \{0, a_1, a_2, \cdots\}.$$

Find the spectrum of T in $B(l^2)$.

3.4 Show that $\sigma(T^*) = \{\bar{\lambda} : \lambda \in \sigma(T)\}$ for every T in $B(H)$.

3.5 If x is invertible in a Banach algebra $\mathcal{A}$, then

$$\sigma(x^{-1}) = \{\lambda^{-1} : \lambda \in \sigma(x)\}.$$

3.6 Let x be an element in a Banach algebra $\mathcal{A}$. Then $r(x^n) = r(x)^n$ for every positive integer n.

3.7 If $\mathcal{B}$ is a maximal abelian subalgebra of a Banach algebra $\mathcal{A}$, then $\mathcal{B}$ is closed and contains the unit $\mathbf{1}$. Also $\sigma_{\mathcal{B}}(x) = \sigma_{\mathcal{A}}(x)$ for all $x \in \mathcal{B}$.

3.8 Show that the spectrum of a matrix A in $M_n(\mathbf{C})$ is the set of eigenvalues of A.

3.9 Show that if A is self-adjoint in $M_n(\mathbf{C})$, then $\|A\| = r(A)$.

3.10 If A is a matrix in $M_n(\mathbf{C})$, then $\|A\|^2 = r(A^*A)$.

3.11 Suppose x is an element of a Banach algebra $\mathcal{A}$. Show that for every open set U in $\mathbf{C}$ that contains $\sigma(x)$ there exists a positive number δ such that $\sigma(y) \subset U$ whenever $y \in \mathcal{A}$ satisfies $\|y - x\| < \delta$.

4

Multiplicative Linear Functionals

4.1 Multiplicative Linear Functionals

In this lecture we study multiplicative linear functionals on a Banach algebra and introduce the notion of the maximal ideal space. When the Banach algebra $\mathcal{A}$ is commutative, we shall show that the set of multiplicative linear functionals on $\mathcal{A}$ can be identified with the set of maximal ideals in $\mathcal{A}$.

DEFINITION 4.1 *A linear functional φ on a Banach algebra $\mathcal{A}$ is multiplicative if φ is nontrivial and $\varphi(xy) = \varphi(x)\varphi(y)$ for all x and y in $\mathcal{A}$.*

Let φ be a multiplicative linear functional on $\mathcal{A}$. It is easy to see that $\varphi(\mathbf{1}) = 1$. In particular, $\|\varphi\| \geq 1$. The next result shows that we always have equality.

PROPOSITION 4.1
Suppose φ is a multiplicative linear functional on a Banach algebra $\mathcal{A}$. Then $\|\varphi\| = 1$.

PROOF Assume $\|\varphi\| > 1$. There must exist a unit vector x in $\mathcal{A}$ such that $|\varphi(x)| > 1$. Let $x_0 = x - \varphi(x)\mathbf{1}$. Clearly $\varphi(x_0) = 0$. On the other hand, the inequality

$$\left\|\mathbf{1} + \frac{x_0}{\varphi(x)}\right\| = \frac{\|x\|}{|\varphi(x)|} < 1$$

and Proposition 2.1 imply that $x_0/\varphi(x)$ (and hence x_0) is invertible. Since φ is multiplicative, we have

$$\varphi(x_0)\varphi(x_0^{-1}) = 1,$$

contradicting the earlier conclusion that $\varphi(x_0) = 0$. ∎

4.2 Examples

EXAMPLE 4.2

If $\mathcal{A} = \mathbf{C}(K)$ for some compact Hausdorff space K, then every point evaluation in K is a multiplicative linear functional on $\mathcal{A}$. We shall see later that these point evaluations are the only multiplicative linear functionals on $\mathbf{C}(K)$. □

EXAMPLE 4.3

If $\mathcal{A} = M_n(\mathbf{C})$ with $n > 1$, there exists no multiplicative linear functional on $\mathcal{A}$. In fact, by the uniqueness of the trace, any multiplicative linear functional on $M_n(\mathbf{C})$ must be a constant multiple of the trace. But it is easy to check that the trace is not multiplicative unless $n = 1$. □

EXAMPLE 4.4

Let $\mathcal{A} = H^\infty(\mathbf{D})$. Every point evaluation in $\mathbf{D}$ is clearly a multiplicative linear functional on $\mathcal{A}$. But the set of multiplicative linear functionals on $H^\infty(\mathbf{D})$ is much larger than the unit disk. See [3][4]. □

4.3 The Maximal Ideal Space

DEFINITION 4.5 *For a Banach algebra $\mathcal{A}$ we let $\mathcal{M}_\mathcal{A}$ denote the set of all multiplicative linear functionals on $\mathcal{A}$. $\mathcal{M}_\mathcal{A}$ is called the maximal ideal space of $\mathcal{A}$ for reasons that will become clear a little later.*

The maximal ideal space of $\mathcal{A}$ is always contained in the closed unit ball of the dual space of the underlying Banach space $\mathcal{A}$. We topologize $\mathcal{M}_\mathcal{A}$ using the weak-star topology inherited from the Banach dual of $\mathcal{A}$.

PROPOSITION 4.2

$\mathcal{M}_\mathcal{A}$ is a compact Hausdorff space with the weak-star topology inherited from the dual space of $\mathcal{A}$.

PROOF It is easy to check that $\mathcal{M}_\mathcal{A}$ is closed in the dual space of $\mathcal{A}$ in the weak-star topology. Since the closed unit ball of the Banach dual of $\mathcal{A}$ is a compact Hausdorff space in the weak-star topology, and any closed subset of a compact Hausdorff space is again compact Hausdorff, we conclude that $\mathcal{M}_\mathcal{A}$ is compact Hausdorff in the weak-star topology. ∎

THEOREM 4.3

If $\mathcal{A}$ is a commutative Banach algebra, then $\mathcal{M}_{\mathcal{A}}$ is in one-to-one correspondence with the set of (proper) maximal ideals in $\mathcal{A}$.

PROOF First assume that φ is a multiplicative linear functional on $\mathcal{A}$. We show that $\ker\varphi$ is a proper maximal ideal in $\mathcal{A}$. In fact, it is clear that $\ker\varphi$ is a proper ideal. To see that it is maximal, let $x \in \mathcal{A} - \ker\varphi$. We can write

$$\mathbf{1} = \left(\mathbf{1} - \frac{x}{\varphi(x)}\right) + \frac{x}{\varphi(x)}.$$

The vector in the parentheses belongs to $\ker\varphi$. We see that the linear span of x and $\ker\varphi$ contains $\mathbf{1}$, and hence any ideal containing both $\ker\varphi$ and x must be the whole algebra. Thus $\ker\varphi$ is maximal.

Next assume that $\mathcal{I}$ is a proper maximal ideal in $\mathcal{A}$. Since no element in $\mathcal{I}$ is invertible, $\|\mathbf{1} - x\| \geq 1$ for all $x \in \mathcal{I}$ by Proposition 2.1. This shows that $\mathbf{1}$ is not in the closure of $\mathcal{I}$. It is easy to see that the closure of $\mathcal{I}$ is again an ideal in $\mathcal{A}$. By the maximality of $\mathcal{I}$ we conclude that $\mathcal{I}$ is equal to its closure and hence $\mathcal{I}$ is closed in $\mathcal{A}$. Since $\mathcal{A}$ is commutative and $\mathcal{I}$ is maximal, the quotient algebra $\mathcal{A}/\mathcal{I}$ is a division algebra (a general fact from abstract algebra). By the Gelfand-Mazur theorem (Theorem 3.2), there exists $\Phi : \mathcal{A}/\mathcal{I} \to \mathbf{C}$ which is an isometric isomorphism. Let π be the quotient mapping from $\mathcal{A}$ onto $\mathcal{A}/\mathcal{I}$, then $\Phi \circ \pi$ is a multiplicative linear functional on $\mathcal{A}$ with kernel $\mathcal{I}$.

Finally, to see that the above correspondence between multiplicative linear functionals and maximal ideals is one-to-one, let φ_1 and φ_2 be multiplicative linear functionals on $\mathcal{A}$ with a common kernel $\mathcal{I}$. We must show that $\varphi_1 = \varphi_2$. For any x in $\mathcal{A}$ we can write

$$\big(\varphi_1(x) - \varphi_2(x)\big)\mathbf{1} = \big(x - \varphi_2(x)\mathbf{1}\big) - \big(x - \varphi_1(x)\mathbf{1}\big).$$

The first term on the right is in $\ker\varphi_2 = \mathcal{I}$ and the second term is in $\ker\varphi_1 = \mathcal{I}$. Thus $\big(\varphi_1(x) - \varphi_2(x)\big)\mathbf{1}$ belongs to $\mathcal{I}$ and we must have $\varphi_1(x) = \varphi_2(x)$, completing the proof of the theorem. ∎

4.4 A Theorem of Gleason, Kahane, and Zelazko

Finally in this lecture we present a characterization of multiplicative linear functionals, a result due to Gleason, Kahane, and Zelazko. In order to prove the theorem we need the following elementary result from complex analysis.

LEMMA 4.4

Suppose $f(z)$ is an entire function in $\mathbf{C}$. If $f(0) = 1$, $f'(0) = 0$, and $0 < |f(z)| \le e^{|z|}$ for all z in $\mathbf{C}$, then $f(z) \equiv 1$.

PROOF Write $f(z) = \exp(g(z))$, where $g(z)$ is entire. It is clear that $g(0) = 0$, $g'(0) = 0$, and $\operatorname{Re} g(z) \le |z|$ for all $z \in \mathbf{C}$. Expanding in terms of real and imaginary parts we see that

$$|g(z)| \le |2r - g(z)|, \qquad |z| \le r,\ r > 0.$$

The function

$$g_r(z) = \frac{r^2 g(z)}{z^2(2r - g(z))}$$

is analytic in $|z| \le r$ and $|g_r(z)| \le 1$ for $|z| = r$. By the maximum principle,

$$|g_r(z)| \le 1, \qquad |z| \le r,\ r > 0.$$

Fix $z \in \mathbf{C}$ and let $r \to +\infty$. We conclude that $g(z) = 0$ or $f(z) = 1$. ∎

THEOREM 4.5

A linear functional φ on a Banach algebra $\mathcal{A}$ is multiplicative if and only if $\varphi(\mathbf{1}) = 1$ and $\varphi(x) \ne 0$ whenever x is invertible in $\mathcal{A}$.

PROOF The "only if" part is clear. To prove the "if" part, assume $\varphi(\mathbf{1}) = 1$ and $\varphi(x) \ne 0$ for every invertible x in $\mathcal{A}$. We proceed to show that $\varphi(xy) = \varphi(x)\varphi(y)$ for all x and y in $\mathcal{A}$.

Let N be the kernel of φ. By assumption N contains no invertible elements of $\mathcal{A}$. By Proposition 2.1 $\|\mathbf{1} - x\| \ge 1$ for all $x \in N$. This easily implies that φ is continuous with norm 1. In fact, for every $x \in \mathcal{A}$ with $\varphi(x) \ne 0$ we can write

$$x = \varphi(x)\left[\mathbf{1} - \left(\mathbf{1} - \frac{x}{\varphi(x)}\right)\right]$$

with

$$\mathbf{1} - \frac{x}{\varphi(x)} \in N,$$

so that

$$\|x\| = |\varphi(x)| \left\|\mathbf{1} - \left(\mathbf{1} - \frac{x}{\varphi(x)}\right)\right\| \ge |\varphi(x)|.$$

Fix $x \in N$ with $\|x\| = 1$ and consider

$$f(z) = \sum_{n=0}^{\infty} \frac{\varphi(x^n)}{n!} z^n, \qquad z \in \mathbf{C}.$$

Since

$$\left|\varphi(x^n)\right| \leq \|x^n\| \leq 1, \qquad n \geq 0,$$

the function f is entire with

$$f(0) = 1, \ f'(0) = 0, \ \left|f(z)\right| \leq e^{|z|}$$

for all $z \in \mathbf{C}$. Also by the continuity of φ we have

$$f(z) = \varphi\left(\sum_{n=0}^{\infty} \frac{(zx)^n}{n!}\right), \qquad z \in \mathbf{C}.$$

It is easy to check that the element

$$\sum_{n=0}^{\infty} \frac{(zx)^n}{n!}$$

is invertible in $\mathcal{A}$ with inverse

$$\sum_{n=0}^{\infty} \frac{(-zx)^n}{n!}.$$

Since N contains no invertible elements, we conclude that f is nonvanishing in $\mathbf{C}$. By Lemma 4.4, $f(z) \equiv 1$. In particular, $\varphi(x^2) = 0$. Thus we have shown that $x^2 \in N$ whenever $x \in N$.

For x and y in $\mathcal{A}$ the assumption $\varphi(\mathbf{1}) = 1$ shows that

$$x = x_1 + \varphi(x)\mathbf{1}, \quad y = y_1 + \varphi(y)\mathbf{1},$$

where x_1 and y_1 belong to N. Therefore,

$$\varphi(xy) = \varphi(x_1 y_1) + \varphi(x)\varphi(y).$$

The proof of the theorem will be completed if we can show that N is closed under multiplication.

Let $x = y$ in the previous paragraph. We obtain

$$\varphi(x^2) = \varphi(x)^2, \qquad x \in \mathcal{A}.$$

Replacing x by $x + y$ leads to

$$\varphi(xy + yx) = 2\varphi(x)\varphi(y), \qquad x, y \in \mathcal{A}.$$

This shows that $xy + yx$ is in N if one of x and y is in N. It follows that the elements

$$(xy + yx)^2, \ x(yxy) + (yxy)x$$

are in N if x is. Since

$$(xy - yx)^2 + (xy + yx)^2 = 2\left[x(yxy) + (yxy)x\right]$$

and

$$(\varphi(xy - yx))^2 = \varphi((xy - yx)^2),$$

we see that $xy - yx$ is in N for all $x \in N$ and $y \in \mathcal{A}$. Adding $xy - yx$ and $xy + yx$, we conclude that xy belongs to N for all x in N and y in $\mathcal{A}$. In particular, N is closed under multiplication, completing the proof of the theorem. ∎

Exercises

4.1 Show that $M_n(\mathbf{C})$ has no nontrivial ideals.

4.2 Every commutative algebra contains proper maximal ideals.

4.3 Convince yourself that there are more multiplicative linear functionals on $H^\infty(\mathbf{D})$ than the point evaluations in $\mathbf{D}$.

4.4 Suppose $\mathcal{A}$ is a Banach algebra whose elements are analytic functions in $\mathbf{D}$ and whose algebraic operations are defined pointwise. Show that $\mathcal{A} \subset H^\infty(\mathbf{D})$ with $\|f\|_\infty \leq \|f\|$ for all f in $\mathcal{A}$, where $\|\ \|$ is the norm in $\mathcal{A}$.

4.5 Can you construct a multiplicative linear functional on the commutative Banach algebra $L^\infty[0, 1]$?

4.6 Are there multiplicative linear functionals on $B(H)$?

4.7 A multiplicative linear functional always maps open sets in $\mathcal{A}$ to open sets in $\mathbf{C}$.

5

The Gelfand Transform and Applications

5.1 The Definition

Recall that the maximal ideal space $\mathcal{M}_{\mathcal{A}}$ of a Banach algebra $\mathcal{A}$ consists of all multiplicative linear functionals on $\mathcal{A}$. The space $\mathcal{M}_{\mathcal{A}}$ is a compact Hausdorff space with the weak-star topology inherited from the Banach dual of $\mathcal{A}$. When $\mathcal{A}$ is commutative, the set $\mathcal{M}_{\mathcal{A}}$ is in one-to-one correspondence with the set of (proper) maximal ideals in $\mathcal{A}$.

DEFINITION 5.1 *Let $\mathcal{A}$ be a Banach algebra. The Gelfand transform is the mapping $\Gamma : \mathcal{A} \to \mathbf{C}(\mathcal{M}_{\mathcal{A}})$ defined by*

$$\Gamma(x)(\varphi) = \varphi(x), \qquad x \in \mathcal{A},\ \varphi \in \mathcal{M}_{\mathcal{A}}.$$

It is easy to see that the Gelfand transform is a contractive Banach algebraic homomorphism from $\mathcal{A}$ into $\mathbf{C}(\mathcal{M}_{\mathcal{A}})$. The Gelfand transform is, in general, neither one-to-one nor onto; the maximal ideal space may even be empty. However, if the Banach algebra $\mathcal{A}$ is commutative, the maximal ideal space is large enough and the Gelfand transform is sensitive enough to distinguish invertible elements. This makes the Gelfand transform a very useful tool in the study of commutative Banach algebras.

5.2 Commutative Banach Algebras

PROPOSITION 5.1

Suppose x is in a commutative Banach algebra $\mathcal{A}$. Then x is invertible in $\mathcal{A}$ if and only if $\Gamma(x)$ is invertible in $\mathbf{C}(\mathcal{M}_{\mathcal{A}})$.

PROOF If x is invertible in $\mathcal{A}$, then $\Gamma(x)$ is clearly invertible in $\mathbf{C}(\mathcal{M}_{\mathcal{A}})$ with inverse $\Gamma(x^{-1})$. If x is not invertible in $\mathcal{A}$, then x is contained in a (proper) maximal ideal $\mathcal{I}$ in A. By Theorem 4.3, there exists $\varphi_0 \in \mathcal{M}_{\mathcal{A}}$ such that $\mathcal{I} = \ker \varphi_0$. It follows that

$$\Gamma(x)(\varphi_0) = \varphi_0(x) = 0,$$

and hence $\Gamma(x)$ is not invertible in $\mathbf{C}(\mathcal{M}_{\mathcal{A}})$. ∎

COROLLARY 5.2

If x is in a commutative Banach algebra $\mathcal{A}$, then

$$\sigma_{\mathcal{A}}(x) = \sigma_{\mathbf{C}(\mathcal{M}_{\mathcal{A}})}\big(\Gamma(x)\big) = \text{Range of } \Gamma(x)$$

and

$$r(x) = \|\Gamma(x)\|_\infty = \sup\big\{|\varphi(x)| : \varphi \in \mathcal{M}_{\mathcal{A}}\big\}.$$

THEOREM 5.3

Suppose $\mathcal{A}$ is a Banach algebra, $x \in \mathcal{A}$, $y \in \mathcal{A}$, and $xy = yx$. Then $\sigma(x+y) \subset \sigma(x) + \sigma(y)$ and $\sigma(xy) \subset \sigma(x)\sigma(y)$.

PROOF By passing to the commutative subalgebra generated by $\mathbf{1}$, x, y, $(\lambda\mathbf{1} - x)^{-1}$, $(\lambda\mathbf{1} - y)^{-1}$, $(\lambda\mathbf{1} - xy)^{-1}$, and $(\lambda\mathbf{1} - x - y)^{-1}$, we may as well assume that $\mathcal{A}$ is commutative. But in this case the desired results follow easily from Corollary 5.2. ∎

5.3 The Spectral Mapping Theorem

Suppose x is in a Banach algebra $\mathcal{A}$ and $f(z)$ is an analytic function in $|z| \le \|x\|$ (so that $f(z)$ is analytic in a neighborhood of $|z| \le \|x\|$). We can define an element $f(x)$ in $\mathcal{A}$ by power series. In fact, if

$$f(z) = \sum_{n=0}^{\infty} a_n z^n$$

is analytic in $|z| \le \|x\|$, then

$$\sum_{n=0}^{\infty} |a_n||z|^n < +\infty$$

for every $|z| \leq \|x\|$. Thus

$$\sum_{n=0}^{\infty} a_n x^n$$

converges to an element in $\mathcal{A}$, which we denote by $f(x)$. Note that the closed disk $|z| \leq \|x\|$ contains the spectrum of x.

THEOREM 5.4 THE SPECTRAL MAPPING THEOREM
Suppose $\mathcal{A}$ is a Banach algebra. If x is in $\mathcal{A}$ and $f(z)$ is analytic in $|z| \leq \|x\|$, then

$$\sigma\big(f(x)\big) = f\big(\sigma(x)\big) = \big\{f(\lambda) : \lambda \in \sigma(x)\big\}.$$

PROOF Fix x in $\mathcal{A}$ and let $\mathcal{A}_0$ be the Banach subalgebra of $\mathcal{A}$ generated by the elements $\mathbf{1}$, x, $(\lambda\mathbf{1} - x)^{-1}$ with $\lambda \in \mathbf{C} - \sigma(x)$, and $\big(\lambda\mathbf{1} - f(x)\big)^{-1}$ with $\lambda \in \mathbf{C} - \sigma\big(f(x)\big)$. Note that for an invertible element y in $\mathcal{A}$ we have $xy = yx$ if and only if $xy^{-1} = y^{-1}x$. We see that $\mathcal{A}_0$ is commutative and

$$\sigma_{\mathcal{A}}(x) = \sigma_{\mathcal{A}_0}(x), \quad \sigma_{\mathcal{A}}\big(f(x)\big) = \sigma_{\mathcal{A}_0}\big(f(x)\big).$$

Let $\Gamma : \mathcal{A}_0 \to \mathbf{C}(\mathcal{M}_{\mathcal{A}_0})$ be the Gelfand transform. By Corollary 5.2

$$\sigma_{\mathcal{A}_0}\big(f(x)\big) = \big\{\varphi\big(f(x)\big) : \varphi \in \mathcal{M}_{\mathcal{A}_0}\big\}.$$

By power series expansion we have

$$\varphi\big(f(x)\big) = f\big(\varphi(x)\big)$$

for each $\varphi \in \mathcal{M}_{\mathcal{A}_0}$. Since

$$\big\{\varphi(x) : \varphi \in \mathcal{M}_{\mathcal{A}_0}\big\} = \sigma_{\mathcal{A}_0}(x)$$

by Corollary 5.2 again, we conclude that

$$\sigma_{\mathcal{A}_0}\big(f(x)\big) = f\big(\sigma_{\mathcal{A}_0}(x)\big).$$

Therefore,

$$\sigma\big(f(x)\big) = f\big(\sigma(x)\big)$$

whenever f is analytic in $|z| \leq \|x\|$. ∎

The spectral mapping theorem can be generalized to the case where f is analytic on $\sigma(x)$ (since $\sigma(x)$ is compact, this means that f is analytic in a neighborhood of $\sigma(x)$). The resulting theorem is called the Riesz functional calculus. See [9].

5.4 The Spectral Radius Formula

Since the spectrum is defined in a purely algebraic way, the spectral radius is a quantity depending only on the algebraic structure of the Banach algebra. However, we shall see a little later that the spectral radius can also be defined in purely topological terms, namely, the norm structure of the Banach algebra.

Suppose f is an analytic function in $|z| < r$ with values in a Banach algebra $\mathcal{A}$. Then we can write

$$f(z) = \sum_{n=0}^{\infty} a_n z^n, \qquad |z| < r,$$

where each a_n belongs to $\mathcal{A}$ and the above series converges uniformly on each compact subset of $|z| < r$. Just as in the scalar case, the coefficients a_n can be computed by a Cauchy type integral formula:

$$a_n = \frac{1}{2\pi i} \int_C \frac{f(\zeta)\, d\zeta}{\zeta^{n+1}}, \qquad n \geq 0,$$

where C is any circle with center 0 and radius less than r. Similarly, if f is an analytic function in $r < |z| < R$ with values in a Banach algebra $\mathcal{A}$, then f admits a Laurent expansion:

$$f(z) = \sum_{n=-\infty}^{+\infty} a_n z^n, \qquad r < |z| < R,$$

where

$$a_n = \frac{1}{2\pi} \int_C \frac{f(\zeta)\, d\zeta}{\zeta^{n+1}}, \qquad -\infty < n < +\infty,$$

and C is any circle with center 0 and radius between r and R.

THEOREM 5.5 THE SPECTRAL RADIUS FORMULA
For every element x in a Banach algebra $\mathcal{A}$ we have

$$r(x) = \lim_{n\to\infty} \|x^n\|^{\frac{1}{n}}.$$

PROOF Since

$$\sigma(x^n) = \{\lambda^n : \lambda \in \sigma(x)\}$$

by the spectral mapping theorem, we have (also see **3.4**)

$$r(x)^n = r(x^n) \leq \|x^n\|, \quad \text{or} \quad r(x) \leq \|x^n\|^{\frac{1}{n}}.$$

It follows that

$$r(x) \leq \liminf_{n\to\infty} \|x^n\|^{\frac{1}{n}}.$$

To prove

$$r(x) \geq \limsup_{n\to\infty} \|x^n\|^{\frac{1}{n}},$$

we consider the domain

$$\Omega = \{z \in \mathbf{C} : |z| > r(x)\}.$$

Since the function

$$\lambda \mapsto (\lambda\mathbf{1} - x)^{-1}$$

is (vector-valued) analytic in Ω, we have a Laurent expansion

$$(\lambda\mathbf{1} - x)^{-1} = \sum_{n=0}^{\infty} \frac{a_n(x)}{\lambda^n}, \qquad |\lambda| > r(x).$$

On the other hand, we know that

$$(\lambda\mathbf{1} - x)^{-1} = \sum_{n=1}^{\infty} \frac{x^{n-1}}{\lambda^n}, \qquad |\lambda| > \|x\|.$$

By the uniqueness of the Laurent expansion in $|\lambda| > \|x\|$, we must have $a_n(x) = x^{n-1}$ for $n \geq 1$ and $a_0(x) = 0$. Therefore,

$$(\lambda\mathbf{1} - x)^{-1} = \sum_{n=1}^{\infty} \frac{x^{n-1}}{\lambda^n}, \qquad |\lambda| > r(x).$$

But it is easy to see that the above series diverges for

$$|\lambda| < \limsup_{n\to\infty} \|x^n\|^{\frac{1}{n}},$$

we conclude that

$$r(x) \geq \limsup_{n\to\infty} \|x^n\|^{\frac{1}{n}},$$

and hence the spectral radius formula holds. ▮

Suppose $\mathcal{A}$ is a Banach algebra and $\mathcal{B}$ is a closed subalgebra containing the unit. If $x \in \mathcal{B}$, then $\sigma_{\mathcal{A}}(x) \neq \sigma_{\mathcal{B}}(x)$ in general. Thus the spectrum of an element depends on the containing algebra. However, the spectral radius formula shows that the spectral radius of an element is independent of the containing Banach algebra; it can be computed using the smallest Banach subalgebra containing x.

5.5 When is the Gelfand Transform Isometric?

THEOREM 5.6
Suppose $\mathcal{A}$ is a commutative Banach algebra and $\Gamma : \mathcal{A} \to \mathbf{C}(\mathcal{M}_{\mathcal{A}})$ is the Gelfand transform. Then Γ is an isometry if and only if $\|x^2\| = \|x\|^2$ for all x in $\mathcal{A}$.

PROOF Recall from Corollary 5.2 that $r(x) = \|\Gamma(x)\|_\infty$. If Γ is an isometry, we then have $r(x) = \|x\|$ for every x in $\mathcal{A}$, and hence

$$\|x^2\| = r(x^2) = r(x)^2 = \|x\|^2$$

for all x in $\mathcal{A}$. Note that the spectral mapping theorem is also used here.

On the other hand, if $\|x^2\| = \|x\|^2$ for every x in $\mathcal{A}$, then by induction

$$\|x^{2^n}\| = \|x\|^{2^n}, \qquad x \in \mathcal{A},\ n \geq 1.$$

By the spectral radius formula, we have $r(x) = \|x\|$ and so $\|\Gamma(x)\|_\infty = \|x\|$ for all x in $\mathcal{A}$. ∎

EXAMPLE 5.2
Let $\mathcal{A} = L^1(\mathbf{Z})$ be the convolution algebra and let $f \in \mathcal{A}$ be defined by

$$f(n) = \begin{cases} 1, & n = 1 \\ -1, & n = 2, 3 \\ 0, & \text{otherwise.} \end{cases}$$

Then $\|f\| = 3$, $\|f\|^2 = 9$, but $\|f^2\| = 7$. Thus by Theorem 5.6 the Gelfand transform of $L^1(\mathbf{Z})$ is not isometric. □

EXAMPLE 5.3
Let $\mathcal{A} = L^\infty$. It is obvious that $\|f\|^2 = \|f^2\|$ for all f in L^∞, so that the Gelfand transform of L^∞ is isometric. We proceed to show that the Gelfand transform of L^∞ is onto. Let $\mathcal{M}$ be the maximal ideal space of L^∞ and let $\mathcal{B}$ be the image of the Gelfand transform. Then $\mathcal{B}$ is closed in $\mathbf{C}(\mathcal{M})$ since the Gelfand transform is isometric. It is easy to see that $\mathcal{B}$ contains all constant functions and $\mathcal{B}$ separates the points of $\mathcal{M}$. By the Stone-Weierstrass approximation theorem $\mathcal{B}$ coincides with $\mathbf{C}(\mathcal{M})$ if we can show that $\mathcal{B}$ is self-adjoint. For f in L^∞ set $f = f_1 + if_2$, where f_1 and f_2 are real-valued. By Corollary 5.2 each of $\Gamma(f_1)$ and $\Gamma(f_2)$ is real-valued. This implies that

$$\overline{\Gamma(f)} = \Gamma(f_1) - i\Gamma(f_2) = \Gamma(\overline{f}),$$

and hence $\mathcal{B}$ is self-adjoint in $\mathbf{C}(\mathcal{M})$. □

EXAMPLE 5.4

Let $\mathcal{A} = H^\infty(\mathbf{D})$. It is clear that $\|f^2\| = \|f\|^2$ for every f in $H^\infty(\mathbf{D})$. Thus the Gelfand transform of $H^\infty(\mathbf{D})$ is isometric. In general, the Gelfand transform of every closed (unital) subalgebra of L^∞ is isometric. The Gelfand transform of such an algebra is onto if and only if the algebra is self-adjoint; see Lecture 8. □

Exercises

5.1 If $xy = yx$ in a Banach algebra $\mathcal{A}$, then $r(xy) \le r(x)r(y)$.

5.2 Show that if $f(z)$ is an analytic function in $\{z : |z| > r\}$ with values in a Banach algebra $\mathcal{A}$ then there exists a sequence $\{a_n\}$ of vectors in $\mathcal{A}$ such that

$$f(z) = \sum_{n=0}^{\infty} a_n z^{-n}$$

for $|z| > r$.

5.3 If $x^2 = x$ in a Banach algebra, then $\sigma(x) \subset \{0, 1\}$.

5.4 If S is a subset of a Banach algebra $\mathcal{A}$, then the centralizer of S, defined by

$$\Lambda(S) = \{x \in \mathcal{A} : xy = yx, y \in S\},$$

is a Banach subalgebra of $\mathcal{A}$.

5.5 Suppose $\mathcal{A}$ is a Banach algebra and S is a subset of $\mathcal{A}$ consisting of mutually commuting elements. Let $\mathcal{B} = \Lambda\big(\Lambda(S)\big)$. Show that $\mathcal{B}$ is commutative and $\sigma_{\mathcal{A}}(x) = \sigma_{\mathcal{B}}(x)$ for every $x \in \mathcal{B}$.

5.6 Let S be a subset of commuting elements in a Banach algebra $\mathcal{A}$. Show that $\Lambda\big(\Lambda(S)\big)$ is the smallest Banach algebra containing S, called the Banach algebra generated by S.

5.7 Show that

$$r(x) = \inf_n \|x^n\|^{\frac{1}{n}}$$

whenever x is an element in a Banach algebra.

6

Examples of Maximal Ideal Spaces

6.1

In this lecture we compute the maximal ideal space for several Banach algebras. By "compute" we mean finding a homeomorphism from the maximal ideal space onto some familar compact Hausdorff space. The Gelfand transform will be determined accordingly.

6.2 The Disc Algebra

Recall that $\mathbf{D} = \{z \in \mathbf{C} : |z| < 1\}$ is the open unit disk and $\overline{\mathbf{D}}$ is the closed unit disk in the complex plane. The disk algebra $A(\mathbf{D})$ is the closed subalgebra of $\mathbf{C}(\overline{\mathbf{D}})$ consisting of functions which are analytic in $\mathbf{D}$. Equivalently, $A(\mathbf{D})$ is the closure in $\mathbf{C}(\overline{\mathbf{D}})$ of the set of polynomials.

For each $z \in \overline{\mathbf{D}}$ we define a multiplicative linear functional $\varphi_z : A(\mathbf{D}) \to \mathbf{C}$ by

$$\varphi_z(f) = f(z), \qquad f \in A(\mathbf{D}).$$

Thus φ_z is the point-evaluation at z. The following result shows that every multiplicative linear functional on $A(\mathbf{D})$ arises in this fashion.

THEOREM 6.1

The mapping $z \mapsto \varphi_z$ is a homeomorphism from $\overline{\mathbf{D}}$ onto the maximal ideal space of the disk algebra $A(\mathbf{D})$.

PROOF Let $\Phi : \overline{\mathbf{D}} \to \mathcal{M}_{A(\mathbf{D})}$ be the mapping defined by $\Phi(z) = \varphi_z$, $z \in \overline{\mathbf{D}}$. It is clear that Φ is one-to-one.

To show that Φ is onto, take $\varphi \in \mathcal{M}_{A(\mathbf{D})}$ and consider the function f_0 in $A(\mathbf{D})$ defined by $f_0(z) = z$, $z \in \overline{\mathbf{D}}$. Then f_0 is a unit vector in $A(\mathbf{D})$. Recall that every multiplicative linear functional has norm 1, so that the point $z_0 = \varphi(f_0)$ lies in $\overline{\mathbf{D}}$. Since φ is linear and multiplicative, we have

$$\varphi(f) = f(z_0) = \varphi_{z_0}(f)$$

for every polynomial f. This implies that $\varphi = \varphi_{z_0}$, because the polynomials are dense in $A(\mathbf{D})$. So Φ is onto.

Suppose $z_\alpha \to z_0$ in $\overline{\mathbf{D}}$. Then $\varphi_{z_\alpha}(f) \to \varphi_{z_0}(f)$ for every f in $A(\mathbf{D})$. By the definition of the weak-star topology in the maximal ideal space of $A(\mathbf{D})$, we have $\varphi_{z_\alpha} \to \varphi_{z_0}$. Thus Φ is a one-to-one continuous mapping from $\overline{\mathbf{D}}$ onto $\mathcal{M}_{A(\mathbf{D})}$. Since both $\overline{\mathbf{D}}$ and $\mathcal{M}_{A(\mathbf{D})}$ are compact Hausdorff, Φ is a homeomorphism. ∎

If we identify the maximal ideal space of $A(\mathbf{D})$ with $\overline{\mathbf{D}}$, then the Gelfand transform is simply the inclusion mapping.

6.3 The Algebra of Continuous Functions

We can apply the techniques used in the proof of Theorem 6.1 to show that the maximal ideal space of $\mathbf{C}(\partial\mathbf{D})$ is homeomorphic to $\partial\mathbf{D}$, where $\partial\mathbf{D}$ is the unit circle in $\mathbf{C}$. This result is a special case of the following general fact.

THEOREM 6.2

Let K be a compact Hausdorff space. For each x in K let $\varphi_x \in \mathcal{M}_{\mathbf{C}(K)}$ be the point evaluation at x. Then the mapping $x \mapsto \varphi_x$ is a homeomorphism from K onto $\mathcal{M}_{\mathbf{C}(K)}$.

PROOF Let Φ denote the mapping $x \mapsto \varphi_x$ from K to $\mathcal{M}_{\mathbf{C}(K)}$. If $x_1 \neq x_2$ in K, it follows from Uryson's lemma that there exists f in $\mathbf{C}(K)$ such that $f(x_1) = 0$ and $f(x_2) = 1$. This shows that the mapping Φ is one-to-one.

To show that Φ is onto, let $\varphi \in \mathcal{M}_{\mathbf{C}(K)}$ and consider

$$\mathcal{I} = \ker\varphi = \{f \in \mathbf{C}(K) : \varphi(f) = 0\}.$$

We first show that there exists x_0 in K such that $f(x_0) = 0$ for all f in $\mathcal{I}$. If this is not the case, then for each x in K there is f_x in $\mathcal{I}$ such that $f_x(x) \neq 0$. By continuity, each x in K has an open neighborhood U_x on which f_x is nonvanishing. By the compactness of K, there exist $x_1, \cdots, x_n$ in K such that

$$K = \bigcup_{k=1}^{n} U_{x_k}.$$

Let

$$f(x) = \sum_{k=1}^{n} |f_{x_k}(x)|^2, \qquad x \in K.$$

Clearly f is nonvanishing on K and hence is invertible in $\mathbf{C}(K)$. This implies that $\varphi(f) \neq 0$. On the other hand, since φ is multiplicative,

$$\varphi(f) = \sum_{k=1}^{n} \varphi(f_{x_k})\varphi(\bar{f}_{x_k}) = 0.$$

This contradiction shows that there must exist some x_0 in K such that $f(x_0) = 0$ for all f in $\mathcal{I}$. Now if f is an arbitrary element in $\mathbf{C}(K)$, $f - \varphi(f)$ is in $\mathcal{I}$, and hence $f(x_0) - \varphi(f) = 0$, or $\varphi(f) = f(x_0)$. Therefore, $\varphi = \Phi(x_0)$ and Φ is onto.

If $x_\alpha \to x$ in K, then $f(x_\alpha) \to f(x)$ for every f in $\mathbf{C}(K)$, and so $\Phi(x_\alpha) \to \Phi(x)$ by the definition of the weak-star topology on $\mathcal{M}_{\mathbf{C}(K)}$. Being a bijective continuous function from the compact Hausdorff space K to the compact Hausdorff space $\mathcal{M}_{\mathbf{C}(K)}$, the mapping Φ must be a homeomorphism. This proves the theorem. ▮

If we identify the maximal ideal space of $\mathbf{C}(K)$ with K, the Gelfand transform is then the identity mapping.

6.4 The Algebra $L^1(\mathbf{Z})$

Recall that $L^1(\mathbf{Z})$ is a commutative Banach algebra with

$$\|f\| = \sum_{n=-\infty}^{\infty} |f(n)|$$

and

$$f * g(n) = \sum_{k=-\infty}^{\infty} f(n-k)g(k).$$

The (multiplicative) unit in $L^1(\mathbf{Z})$ is the function $\mathbf{1}$ in $L^1(\mathbf{Z})$ defined by $\mathbf{1}(0) = 1$ and $\mathbf{1}(n) = 0$ for $n \in \mathbf{Z} - \{0\}$.

THEOREM 6.3

The maximal ideal space of the Banach algebra $L^1(\mathbf{Z})$ is homeomorphic to the unit circle $\partial\mathbf{D}$.

PROOF For each $z \in \partial\mathbf{D}$ define $\varphi_z : L^1(\mathbf{Z}) \to \mathbf{C}$ by

$$\varphi_z(f) = \sum_{n=-\infty}^{\infty} f(n)z^n, \qquad f \in L^1(\mathbf{Z}).$$

The functional φ_z is clearly linear, bounded, and nontrivial. Moreover, by Fubini's theorem

$$\begin{aligned}\varphi_z(f * g) &= \sum_{n=-\infty}^{\infty} z^n \sum_{k=-\infty}^{\infty} f(n-k)g(k) \\ &= \sum_{k=-\infty}^{\infty} g(k)z^k \sum_{n=-\infty}^{\infty} f(n-k)z^{n-k} \\ &= \sum_{k=-\infty}^{\infty} g(k)z^k \sum_{n=-\infty}^{\infty} f(n)z^n \\ &= \varphi_z(f)\varphi_z(g).\end{aligned}$$

Thus φ_z is multiplicative and hence is in the maximal ideal space of $L^1(\mathbf{Z})$.

Define $\Phi : \partial\mathbf{D} \to \mathcal{M}_{L^1(\mathbf{Z})}$ by $\Phi(z) = \varphi_z$. We show that Φ is a surjective homeomorphism.

For each n in $\mathbf{Z}$ let f_n be the function in $L^1(\mathbf{Z})$ defined by $f_n(n) = 1$ and $f_n(k) = 0$ for $k \neq n$. It is easy to check the following elementary properties: $f_0 = \mathbf{1}$, $\|f_n\| = 1$, and $f_n * f_k = f_{n+k}$ for all n and k in $\mathbf{Z}$.

That Φ is one-to-one follows from the observation that $\Phi(z)(f_1) = z$ for all z in $\partial\mathbf{D}$.

To see that Φ is onto, let $\varphi \in \mathcal{M}_{L^1(\mathbf{Z})}$ and let $z_0 = \varphi(f_1)$. Clearly $|z_0| \leq 1$. On the other hand, $\mathbf{1} = f_1 * f_{-1}$ implies that $1 = z_0\varphi(f_{-1})$ and hence $1 = |z_0|\,|\varphi(f_{-1})| \leq |z_0|$. Thus z_0 is in $\partial\mathbf{D}$. Since φ is multiplicative,

$$\varphi(f_n) = \varphi(f_1 * \cdots * f_1) = \varphi(f_1)^n = z_0^n$$

for all positive integers n. It follows from this and the identity $\mathbf{1} = f_n * f_{-n}$ that $\varphi(f_n) = z_0^n$ for all integers n. If f is in $L^1(\mathbf{Z})$, we can write

$$f = \sum_{n=-\infty}^{\infty} f(n)f_n$$

with the series converging in $L^1(\mathbf{Z})$. By the continuity and linearity of φ, we have

$$\varphi(f) = \sum_{n=-\infty}^{\infty} f(n)\varphi(f_n) = \sum_{n=-\infty}^{\infty} f(n)z_0^n = \varphi_{z_0}(f)$$

for all f in $L^1(\mathbf{Z})$ and hence Φ is onto.

If $z_\alpha \to z$ in $\partial\mathbf{D}$, then by an easy $\epsilon - \delta$ argument

$$\lim_{\alpha} \sum_{n=-\infty}^{\infty} f(n) z_\alpha^n = \sum_{n=-\infty}^{\infty} f(n) z^n$$

for every f in $L^1(\mathbf{Z})$ and hence $\Phi(z_\alpha) \to \Phi(z)$, showing that Φ is continuous. Again since both $\partial\mathbf{D}$ and $\mathcal{M}_{L^1(\mathbf{Z})}$ are compact Hausdorff, Φ must be a homeomorphism. ∎

If we identify the maximal ideal space of $L^1(\mathbf{Z})$ with the unit circle $\partial\mathbf{D}$, the Gelfand transform $\Gamma : L^1(\mathbf{Z}) \to \mathbf{C}(\partial\mathbf{D})$ takes the following form:

$$\Gamma(f)(z) = \sum_{n=-\infty}^{\infty} f(n) z^n, \quad f \in L^1(\mathbf{Z}),\ z \in \partial\mathbf{D}.$$

For a function f in $\mathbf{C}(\partial\mathbf{D})$ its Fourier coefficients are given by

$$a_n = \frac{1}{2\pi} \int_0^{2\pi} f(e^{i\theta}) e^{-in\theta}\, d\theta, \qquad n \in \mathbf{Z},$$

so that

$$f(z) = \sum_{n=-\infty}^{\infty} a_n z^n, \qquad z \in \partial\mathbf{D}.$$

It is then clear that a function f in $\mathbf{C}(\partial\mathbf{D})$ is in the range of the Gelfand transform $\Gamma : L^1(\mathbf{Z}) \to \mathbf{C}(\partial\mathbf{D})$ if and only if its Fourier coefficients are absolutely summable. Since

$$f(z) = \sum_{n=1}^{\infty} \frac{e^{in \log n}}{n} z^n$$

is a function in $\mathbf{C}(\partial\mathbf{D})$ (see [12]) whose Fourier coefficients are not absolutely summable, the Gelfand transform $\Gamma : L^1(\mathbf{Z}) \to \mathbf{C}(\partial\mathbf{D})$ is not onto. Recall that the Gelfand transform is not isometric either.

6.5 The Algebra of Bounded Continuous Functions

Let Ω be a locally compact (but noncompact) Hausdorff space and let $BC(\Omega)$ be the algebra of all bounded complex-valued continuous functions on Ω. It is easy to see that $BC(\Omega)$ is a Banach algebra with the sup-norm and pointwise algebraic operations.

A compact Hausdorff space K is called the Stone-Čech compactification of Ω if K satisfies the following conditions: There exists a homeomorphism Φ from

Ω onto a dense subset $\widetilde{\Omega}$ of K such that for every f in $BC(\Omega)$ the function $f \circ \Phi^{-1} : \widetilde{\Omega} \to \mathbf{C}$ extends to a continuous function on K. It is well known that the Stone-Čech compactification of Ω is unique up to homeomorphism. We use the symbol $\beta\Omega$ to denote the Stone-Čech compactification of Ω.

THEOREM 6.4

The maximal ideal space of $BC(\Omega)$ is homeomorphic to the the Stone-Čech compactification $\beta\Omega$ of Ω.

PROOF Let $\mathcal{M}$ be the maximal ideal space of $BC(\Omega)$. Define $\Phi : \Omega \to \mathcal{M}$ by

$$\Phi(x)(f) = f(x), \qquad x \in \Omega, \ f \in BC(\Omega).$$

Namely, $\Phi(x)$ is the point-evaluation at $x \in \Omega$. It is clear that $x_\alpha \to x$ in Ω if and only if $f(x_\alpha) \to f(x)$ for all f in $BC(\Omega)$ if and only if $\Phi(x_\alpha) \to \Phi(x)$ in $\mathcal{M}$. Thus Φ is a homeomorphism from Ω onto a subset $\widetilde{\Omega}$ of $\mathcal{M}$. For every f in $BC(\Omega)$ the function $f \circ \Phi^{-1} : \widetilde{\Omega} \to \mathbf{C}$ extends to a continuous function on $\mathcal{M}$, namely, the function $\Gamma(f)$, where Γ is the Gelfand transform of $BC(\Omega)$. It remains to show that $\widetilde{\Omega}$ is dense in $\mathcal{M}$.

Assume $\widetilde{\Omega}$ is not dense in $\mathcal{M}$. Then there exists φ_0 in $\mathcal{M}$ and an open neighborhood U of φ_0 in $\mathcal{M}$ such that $U \cap \widetilde{\Omega} = \emptyset$. Since $\mathcal{M}$ is topologized with the weak-star topology, we may as well assume that

$$U = \Big\{\varphi \in \mathcal{M} : \big|\varphi(f_k) - \varphi_0(f_k)\big| < \epsilon, 1 \le k \le n\Big\},$$

where ϵ is a positive number and $f_1, \cdots, f_n$ are functions in $BC(\Omega)$. Consider the function

$$f(x) = \sum_{k=1}^{n} \big|f_k(x) - \varphi_0(f_k)\big|^2, \qquad x \in \Omega.$$

That $\widetilde{\Omega} \cap U = \emptyset$ implies that $f(x) \ge \epsilon^2$ for every x in Ω. In particular, f is invertible in $BC(\Omega)$, so that $\varphi_0(f) \neq 0$. On the other hand,

$$\varphi_0(f) = \sum_{k=1}^{n} \varphi_0\big(f_k - \varphi_0(f_k)\big)\varphi_0\big(\bar{f}_k - \overline{\varphi_0(f_k)}\,\big) = 0.$$

This contradiction shows that $\widetilde{\Omega}$ is dense in $\mathcal{M}$, and hence $\mathcal{M}$ is homeomorphic to $\beta\Omega$. ∎

If we think of Ω as a subset of the maximal ideal space of $BC(\Omega)$, then the Gelfand transform of $f \in BC(\Omega)$, $\Gamma(f)$, is simply the unique extension of f to $\mathcal{M}$.

Exercises

6.1 If f in $\mathbf{C}(\partial\mathbf{D})$ has an absolutely convergent Fourier series and f is non-vanishing on $\partial\mathbf{D}$, then $1/f$ has an absolutely convergent Fourier series. This result is called Wiener's theorem in Fourier analysis.

6.2 Use the method in **6.2** to show that the maximal ideal space of $\mathbf{C}(\partial\mathbf{D})$ is homeomorphic to $\partial\mathbf{D}$.

6.3 Let
$$\mathcal{A} = \{f \in \mathbf{C}[0,1] : f' \in \mathbf{C}[0,1]\}$$
with pointwise operations and
$$\|f\| = \|f\|_\infty + \|f'\|_\infty.$$
Show that the maximal ideal space of $\mathcal{A}$ is homeomorphic to $[0,1]$. The Gelfand transform for $\mathcal{A}$ is neither isometric nor onto.

6.4 Show that the abstract index group of $L^1(\mathbf{Z})$ is isomorphic to $\mathbf{Z}$.

6.5 Let B_n be the open unit ball in $\mathbf{C}^n$ and let $A(B_n)$ be the "ball algebra" consisting of holomorphic functions in B_n which can be extended continuously to the closed unit ball $\bar{B}_n$. It is clear that $A(B_n)$ is a Banach algebra with the sup-norm and pointwise operations. Show that the maximal ideal space of $A(B_n)$ can be identified with $\bar{B}_n$ via point-evaluations.

6.6 Show that the spectrum of f in $L^1(\mathbf{Z})$ is the range of the function
$$\sum_{n=-\infty}^{+\infty} f(n)z^n, \qquad |z| = 1.$$
Thus f is invertible in $L^1(\mathbf{Z})$ if and only if
$$\sum_{n=-\infty}^{+\infty} f(n)z^n \neq 0, \qquad |z| = 1,$$
and the spectral radius of f is
$$r(f) = \sup\left\{\left|\sum_{n=-\infty}^{+\infty} f(n)z^n\right| : |z| = 1\right\}.$$

6.7 Suppose $\mathcal{A}$ is a Banach algebra and x is a nonzero element in $\mathcal{A}$. Let $\mathcal{B}$ be the closed subalgebra generated **1** and x. Show that $\mathcal{B}$ is commutative and the maximal ideal space of $\mathcal{B}$ is homeomorphic to $\sigma_{\mathcal{B}}(x)$.

7

Non-Unital Banach Algebras

7.1 The Unitization

For some applications it may be necessary to consider non-unital Banach algebras. However, our approach in the book does not lose much generality. The following argument shows that every (non-unital) Banach algebra can be thought of as a closed subalgebra of a unital Banach algebra.

Let $\mathcal{A}$ be a complex algebra (possibly without unit) equipped with a complete norm $\| \ \|$ satisfying $\|xy\| \leq \|x\| \, \|y\|$ for all x and y in $\mathcal{A}$. Let $\widetilde{\mathcal{A}}$ be the set of all ordered pairs (x, a), where $x \in \mathcal{A}$ and $a \in \mathbf{C}$. $\widetilde{\mathcal{A}}$ is a vector space with linear operations defined componentwise. If we define a multiplication in $\widetilde{\mathcal{A}}$ by

$$(x, a)(y, b) = (xy + ay + bx, ab)$$

and define a norm in $\widetilde{\mathcal{A}}$ by

$$\big\|(x, a)\big\| = \|x\| + |a|,$$

then it is easy to check that $\widetilde{\mathcal{A}}$ becomes a Banach algebra with unit $\mathbf{1} = (0, 1)$. It is obvious that the mapping $x \mapsto (x, 0)$ is an isometric isomorphism from $\mathcal{A}$ onto a closed subspace (actually a closed two-sided ideal) of $\widetilde{\mathcal{A}}$. We shall call $\widetilde{\mathcal{A}}$ the unitization of $\mathcal{A}$.

7.2 Two Examples

The following are two classical examples of non-unital commutative Banach algebras.

EXAMPLE 7.1

Let Ω be a locally compact (but not compact) Hausdorff space. Let $\mathbf{C}_0(\Omega)$ be the space of complex-valued functions on Ω that can be uniformly approximated by continuous functions on Ω with compact support. The space $\mathbf{C}_0(\Omega)$ is sometimes referred to as the space of continuous functions on Ω which vanish at ∞, because a continuous function f on Ω belongs to $\mathbf{C}_0(\Omega)$ if and only if for every $\epsilon > 0$ there exists a compact set Ω_ϵ in Ω such that $|f(x)| < \epsilon$ for all $x \in \Omega - \Omega_\epsilon$. If $\widetilde{\Omega} = \Omega \cup \{\infty\}$ is the one-point compactification of Ω, then a function f in $\mathbf{C}(\widetilde{\Omega})$ belongs to $\mathbf{C}_0(\Omega)$ if and only if $f(\infty) = 0$. The space $\mathbf{C}_0(\Omega)$ is a non-unital commutative Banach algebra when endowed with the sup-norm and pointwise algebraic operations. □

EXAMPLE 7.2

The Banach space $L^1(\mathbf{R}, dx)$ becomes a non-unital commutative Banach algebra when multiplication is defined as convolution:

$$f * g(x) = \int_{-\infty}^{\infty} f(x-y)g(y)\,dy, \qquad x \in \mathbf{R}.$$

Similarly, if G is a locally compact (but not compact) abelian group and if dm is the Haar measure on G, then the Banach space $L^1(G, dm)$ is a commutative Banach algebra without identity when multiplication is defined as convolution:

$$f * g(x) = \int_G f(xy^{-1})g(y)\,dm(y), \qquad x \in G.$$

In fact, by Fubini's theorem

$$\begin{aligned}\int_G |f * g(x)|\,dm(x) &\leq \int_G dm(x) \int_G |f(xy^{-1})g(y)|\,dm(y) \\ &= \int_G |g(y)|\,dm(y) \int_G |f(xy^{-1})|\,dm(x) \\ &= \int_G |g(y)|\,dm(y) \int_G |f(x)|\,dm(x)\end{aligned}$$

since dm is invariant under translation. □

7.3 The Maximal Ideal Space

Let $\mathcal{A}$ be a Banach algebra without identity. The maximal ideal space $\mathcal{M}_\mathcal{A}$ can be defined exactly as before. Thus $\mathcal{M}_\mathcal{A}$ consists of nonzero multiplicative linear functionals in $\mathcal{A}$.

PROPOSITION 7.1

If $\mathcal{A}$ is a non-unital Banach algebra and $\varphi : \mathcal{A} \to \mathbf{C}$ is a multiplicative linear functional on $\mathcal{A}$, then $\left|\varphi(x)\right| \leq \|x\|$ for all x in $\mathcal{A}$.

PROOF Let $\widetilde{\mathcal{A}} = \left\{(x, a) : x \in \mathcal{A}, a \in \mathbf{C}\right\}$ be the unitization of $\mathcal{A}$ described in **7.1**. If φ is a multiplicative linear functional on $\mathcal{A}$, we can extend it to $\widetilde{\varphi} : \widetilde{\mathcal{A}} \to \mathbf{C}$ as follows:

$$\widetilde{\varphi}(x, a) = \varphi(x) + a, \qquad x \in \mathcal{A},\ a \in \mathbf{C}.$$

It is easy to see that $\widetilde{\varphi}$ as defined above is multiplicative on $\widetilde{\mathcal{A}}$. By Proposition 4.1

$$\left|\widetilde{\varphi}(x, a)\right| \leq \left\|(x, a)\right\| = \|x\| + |a|, \qquad x \in \mathcal{A},\ a \in \mathbf{C}.$$

Letting $a = 0$ yields $\left|\varphi(x)\right| \leq \|x\|$ for all $x \in \mathcal{A}$. ∎

THEOREM 7.2

The maximal ideal space of a non-unital Banach algebra $\mathcal{A}$, endowed with the weak-star topology of the Banach dual $\mathcal{A}^*$ of $\mathcal{A}$, is a locally compact Hausdorff space.

PROOF By Proposition 7.1, the maximal ideal space $\mathcal{M}_{\mathcal{A}}$ of $\mathcal{A}$ is contained in the closed unit ball of $\mathcal{A}^*$. It is easy to see that $\mathcal{M}_{\mathcal{A}} \cup \{0\}$ is closed, and hence compact by Alaoglu's theorem, in the weak-star topology of $\mathcal{A}^*$. This shows that $\mathcal{M}_{\mathcal{A}}$ is locally compact. That $\mathcal{M}_{\mathcal{A}}$ is Hausdorff is clear. ∎

If $\mathcal{A}$ is unital and $\{\varphi_\alpha\}$ is a net in $\mathcal{M}_{\mathcal{A}}$ converging to φ in $\mathcal{A}^*$, then φ is multiplicative and $\varphi(\mathbf{1}) = \lim_\alpha \varphi_\alpha(\mathbf{1}) = 1$, so that φ is nontrivial and $\varphi \in \mathcal{M}_{\mathcal{A}}$. Thus $\mathcal{M}_{\mathcal{A}}$ itself is compact. In the non-unital case, however, we can only conclude that $\mathcal{M}_{\mathcal{A}} \cup \{0\}$ is compact, because the limit of a net of nonzero multiplicative linear functionals on $\mathcal{A}$ may converge to 0.

7.4 The Gelfand Transform

Let $\mathcal{M}_{\mathcal{A}}$ be the maximal ideal space of a non-unital Banach algebra $\mathcal{A}$. The Gelfand transform $\Gamma : \mathcal{A} \to \mathbf{C}(\mathcal{M}_{\mathcal{A}})$ can be defined exactly as in the unital case. Here $\mathbf{C}(\mathcal{M}_{\mathcal{A}})$ is the space of all complex-valued continuous functions in $\mathcal{M}_{\mathcal{A}}$. The next result shows that the image of the Gelfand transform is actually contained in the non-unital Banach algebra $\mathbf{C}_0(\mathcal{M}_{\mathcal{A}})$.

THEOREM 7.3
Suppose φ is a non-unital Banach algebra. Then the Gelfand transform is a contractive homomorphism from $\mathcal{A}$ into $\mathbf{C}_0(\mathcal{M}_{\mathcal{A}})$.

PROOF That Γ is a contractive homomorphism is clear. For $x \in \mathcal{A}$ and $\epsilon > 0$ the set

$$\Omega_\epsilon = \{\varphi \in \mathcal{M}_{\mathcal{A}} : |\varphi(x)| = |\Gamma(x)(\varphi)| \geq \epsilon\}$$

is weak-star closed, and hence compact, in $\mathcal{M}_{\mathcal{A}}$. Thus for any $\epsilon > 0$ there exists a compact set Ω_ϵ in $\mathcal{M}_{\mathcal{A}}$ such that $|\Gamma(x)(\varphi)| < \epsilon$ for all $x \in \mathcal{M}_{\mathcal{A}} - \Omega_\epsilon$. This shows that $\Gamma(x)$ vanishes at ∞ and so $\Gamma(x) \in \mathbf{C}_0(\mathcal{M}_{\mathcal{A}})$. ∎

7.5 The Algebra $\mathbf{C}_0(\Omega)$

We now identify the maximal ideal space of $\mathbf{C}_0(\Omega)$ and the corresponding Gelfand transform.

THEOREM 7.4
If Ω is a locally compact (but not compact) Hausdorff space, then the maximal ideal space $\mathcal{M}$ of $\mathbf{C}_0(\Omega)$ is homeomorphic to Ω via point-evaluations. The Gelfand transform of $\mathbf{C}_0(\Omega)$, with this identification of $\mathcal{M}$, is simply the identity mapping.

PROOF It is clear that every point-evaluation in Ω gives rise to a nontrivial multiplicative linear functional on $\mathbf{C}_0(\Omega)$. Conversely, if φ is a nontrivial multiplicative linear functional on $\mathbf{C}_0(\Omega)$, we show that φ arises as a point-evaluation in Ω.

Let $\widetilde{\Omega} = \Omega \cup \{0\}$ be the one-point compactification of Ω. Thus a function in $\mathbf{C}(\widetilde{\Omega})$ belongs to $\mathbf{C}_0(\Omega)$ if and only if $f(0) = 0$. Define $\widetilde{\varphi} : \mathbf{C}(\widetilde{\Omega}) \to \mathbf{C}$ by

$$\widetilde{\varphi}(f) = \varphi(f - f(0)) + f(0), \qquad f \in \mathbf{C}(\widetilde{\Omega}).$$

It is clear that $\widetilde{\varphi}$ is linear. We show that $\widetilde{\varphi}$ is also multiplicative.

Given f and g in $\mathbf{C}(\widetilde{\Omega})$ we write

$$fg = (f - f(0))(g - g(0)) + f(0)(g - g(0)) + g(0)(f - f(0)) + f(0)g(0).$$

Then

$$\begin{aligned}\widetilde{\varphi}(fg) &= \varphi(f - f(0))\varphi(g - g(0)) + f(0)\varphi(g - g(0)) \\ &\quad + g(0)\varphi(f - f(0)) + f(0)g(0) \\ &= \left[\varphi(f - f(0)) + f(0)\right]\left[\varphi(g - g(0)) + g(0)\right] \\ &= \widetilde{\varphi}(f)\widetilde{\varphi}(g).\end{aligned}$$

Thus $\widetilde{\varphi}$ is a nontrivial multiplicative linear functional on $\mathbf{C}(\widetilde{\Omega})$. By Theorem 6.2 there exists a point $\lambda \in \Omega \cup \{0\}$ such that

$$\widetilde{\varphi}(f) = \varphi(f - f(0)) + f(0) = f(\lambda)$$

for all f in $\mathbf{C}(\widetilde{\Omega})$. In particular, $\varphi(f) = f(\lambda)$ for all f in $\mathbf{C}_0(\Omega)$. Since φ is nontrivial, we must have $\lambda \in \Omega$.

Define $\Phi : \Omega \to \mathcal{M}$ by

$$\Phi(\lambda)(f) = f(\lambda), \qquad \lambda \in \Omega, f \in \mathbf{C}_0(\Omega).$$

We have shown that Φ is onto. It is easy to see that Φ is one-to-one. Also, $\lambda_\alpha \to \lambda$ in Ω if and only if $f(\lambda_\alpha) \to f(\lambda)$ for every $f \in \mathbf{C}_0(\Omega)$ if and only if $\Phi(\lambda_\alpha) \to \Phi(\lambda)$ in $\mathcal{M}$, so that Φ is a homeomorphism from Ω onto $\mathcal{M}$.

If we identify the maximal ideal space of $\mathbf{C}_0(\Omega)$ with Ω as above, it is clear that the Gelfand transform is the identity mapping. ▮

7.6 The Algebra $L^1(\mathbf{R}, dx)$

We proceed to identify the maximal ideal space of the non-unital Banach algebra $L^1(\mathbf{R}, dx)$. The corresponding Gelfand transform will be shown to coincide with the familiar Fourier transform.

LEMMA 7.5

For every $t \in \mathbf{R}$ the linear functional φ_t on $L^1(\mathbf{R}, dx)$, defined by

$$\varphi_t(f) = \int_{\mathbf{R}} f(x) e^{itx}\, dx, \qquad f \in L^1(\mathbf{R}, dx),$$

is multiplicative.

PROOF Suppose f and g are in $L^1(\mathbf{R}, dx)$. By Fubini's theorem and the invariance of dx under translations

$$\begin{aligned}
\varphi_t(f * g) &= \int_{\mathbf{R}} f * g(x) e^{itx}\, dx \\
&= \int_{\mathbf{R}} e^{itx}\, dx \int_{\mathbf{R}} f(x-y) g(y)\, dy \\
&= \int_{\mathbf{R}} e^{ity} g(y)\, dy \int_{\mathbf{R}} e^{it(x-y)} f(x-y)\, dx \\
&= \int_{\mathbf{R}} e^{ity} g(y)\, dy \int_{\mathbf{R}} e^{itx} f(x)\, dx \\
&= \varphi_t(f)\varphi_t(g),
\end{aligned}$$

so that φ_t is multiplicative. ∎

LEMMA 7.6

Suppose $f : \mathbf{R} \to \mathbf{C}$ is continuous and bounded. If $f(x)f(y) = f(x+y)$ for all $x, y \in \mathbf{R}$, then either $f \equiv 0$ or there exists $t \in \mathbf{R}$ such that $f(x) = e^{itx}$ for all $x \in \mathbf{R}$.

PROOF Suppose f is not identically zero. Then it is clear that $f(0) = 1$. By the continuity of f there exists $\delta > 0$ such that

$$c = \int_0^\delta f(y)\,dy \neq 0.$$

For every $x \in \mathbf{R}$ we have

$$cf(x) = \int_0^\delta f(x)f(y)\,dy = \int_0^\delta f(x+y)\,dx = \int_x^{\delta+x} f(y)\,dy.$$

Since f is continuous, the last integral above is a differentiable function of x, so that f is continuously differentiable. Differentiating the equation $f(x+y) = f(x)f(y)$ with respect to y, then setting $y = 0$, we obtain $f'(x) = \lambda f(x)$, where $\lambda = f'(0)$. This, along with the initial condition that $f(0) = 1$, shows that $f(x) = e^{\lambda x}$ for all x in $\mathbf{R}$. The boundedness of f implies that λ must be purely imaginary. ∎

THEOREM 7.7

The mapping $t \mapsto \varphi_t$ is a homeomorphism from $\mathbf{R}$ onto the maximal ideal space of $L^1(\mathbf{R}, dx)$.

PROOF Write $\Phi(t) = \varphi_t$, $t \in \mathbf{R}$. By Lemma 7.5, Φ maps $\mathbf{R}$ into $\mathcal{M}$, the maximal ideal space of $L^1(\mathbf{R}, dx)$. We next show that Φ is onto.

Suppose $\varphi \in \mathcal{M}$. In particular, φ is in the Banach dual of $L^1(\mathbf{R}, dx)$, so that there exists a function $h \in L^\infty(\mathbf{R}, dx)$ such that

$$\varphi(f) = \int_{\mathbf{R}} f(x)h(x)\,dx, \qquad f \in L^1(\mathbf{R}, dx).$$

If f and g are in $L^1(\mathbf{R}, dx)$, then by Fubini's theorem

$$\begin{aligned}
\varphi(f * g) &= \int_{\mathbf{R}} f * g(x)h(x)\,dx \\
&= \int_{\mathbf{R}} h(x)\,dx \int_{\mathbf{R}} f(x-y)g(y)\,dy \\
&= \int_{\mathbf{R}} g(y)\,dy \int_{\mathbf{R}} f_y(x)h(x)\,dx \\
&= \int_{\mathbf{R}} g(y)\varphi(f_y)\,dy,
\end{aligned}$$

where $f_y(x) = f(x-y)$. On the other hand, for all f and g in $L^1(\mathbf{R}, dx)$

$$\varphi(f * g) = \varphi(f)\varphi(g) = \varphi(f) \int_{\mathbf{R}} g(y)h(y)\,dy.$$

We conclude that $\varphi(f)h(y) = \varphi(f_y)$ for every f in $L^1(\mathbf{R}, dx)$ and almost all $y \in \mathbf{R}$. Fix $f \in L^1(\mathbf{R}, dx)$ so that $\varphi(f) \neq 0$. Since $y \mapsto f_y$ is a continuous mapping from $\mathbf{R}$ into $L^1(\mathbf{R}, dx)$, the function $h(y) = \varphi(f_y)/\varphi(f)$ is continuous.

If we replace y by $x+y$ and then f by f_x in the equation $\varphi(f)h(y) = \varphi(f_y)$, we obtain

$$\varphi(f)h(x+y) = \varphi(f)h(x)h(y),$$

so that $h(x+y) = h(x)h(y)$ for all x and y in $\mathbf{R}$. By Lemma 7.6, there exists $t \in \mathbf{R}$ such that $h(x) = e^{itx}$ for almost every $x \in \mathbf{R}$. Therefore, $\varphi = \varphi_t$ and Φ is onto.

If $t_\alpha \to t$ in $\mathbf{R}$, an easy $\epsilon - \delta$ argument shows that $\varphi_{t_\alpha}(f) \to \varphi_t(f)$ for every $f \in L^1(\mathbf{R}, dx)$, so that $\Phi(t_\alpha) \to \Phi(t)$ in $\mathcal{M}$. On the other hand, if

$$\lim_\alpha \int_{\mathbf{R}} f(x)\big(e^{it_\alpha x} - e^{itx}\big)\,dx = 0$$

for every f in $L^1(\mathbf{R}, dx)$, then by letting

$$f(x) = \begin{cases} e^{-itx} & 0 \le x \le 1 \\ 0 & \text{otherwise} \end{cases}$$

one can show that $t_\alpha \to t$ in $\mathbf{R}$. We conclude that Φ is a homeomorphism from $\mathbf{R}$ onto $\mathcal{M}$. ∎

COROLLARY 7.8

Suppose $h \in L^\infty(\mathbf{R}, dx)$ and $h(x-y)h(y) = h(x)$ almost everywhere on $\mathbf{R} \times \mathbf{R}$. Then either $h = 0$ or there exists $t \in \mathbf{R}$ such that $h(x) = e^{itx}$ for almost all x.

PROOF Define $\varphi : L^1(\mathbf{R}, dx) \to \mathbf{C}$ by

$$\varphi(f) = \int_{\mathbf{R}} f(x)h(x)\,dx, \qquad f \in L^1(\mathbf{R}, dx).$$

It is clear that φ is a nonzero bounded linear functional on $L^1(\mathbf{R}, dx)$, provided that $h \neq 0$. Furthermore, if f and g are in $L^1(\mathbf{R}, dx)$, then by Fubini's theorem

$$\begin{aligned} 0 &= \int_{\mathbf{R}} \int_{\mathbf{R}} f(x-y)g(y)\big(h(x-y)h(y) - h(x)\big)\,dy\,dx \\ &= \int_{\mathbf{R}} h(y)g(y)\,dy \int_{\mathbf{R}} f(x-y)h(x-y)\,dx \\ &\quad - \int_{\mathbf{R}} h(x)\,dx \int_{\mathbf{R}} f(x-y)g(y)\,dy \end{aligned}$$

$$
\begin{aligned}
&= \int_{\mathbf{R}} g(y)h(y)\,dy \int_{\mathbf{R}} f(x)h(x)\,dx - \int_{\mathbf{R}} f * g(x)h(x)\,dx \\
&= \varphi(f)\varphi(g) - \varphi(f * g),
\end{aligned}
$$

so that φ is multiplicative. By Theorem 7.7, there exists $t \in \mathbf{R}$ such that

$$\varphi(f) = \int_{\mathbf{R}} f(x)h(x)\,dx = \int_{\mathbf{R}} f(x)e^{itx}\,dx$$

for all f in $L^1(\mathbf{R}, dx)$. This shows that $h(x) = e^{itx}$ for almost all x in $\mathbf{R}$, provided that h is nonzero in $L^\infty(\mathbf{R})$. ∎

REMARK 7.9 If we identify the maximal ideal space $\mathcal{M}$ of $L^1(\mathbf{R}, dx)$ with $\mathbf{R}$, the Gelfand transform is then the well-known Fourier transform on $\mathbf{R}$. ∎

The results above about $L^1(\mathbf{R}, dx)$ can be generalized to $L^1(G, dm)$, where G is a locally compact abelian group and dm is the Haar measure on G. In this case, the maximal ideal space of $L^1(G, dm)$ can be identified with the dual group $\widehat{G}$ of G, and the Gelfand transform is the generalized Fourier transform. See [10].

Part II

C^*-ALGEBRAS

8

C^*-Algebras

8.1 The Definition

DEFINITION 8.1 *A C^*-algebra is a Banach algebra $\mathcal{A}$ together with a mapping $x \mapsto x^*$ on $\mathcal{A}$ satisfying the following conditions:*

(a) $(x^*)^* = x$ *for all* $x \in \mathcal{A}$.

(b) $(ax + by)^* = \bar{a}x^* + \bar{b}y^*$ *for all* $x, y \in \mathcal{A}$ *and* $a, b \in \mathbf{C}$.

(c) $(xy)^* = y^*x^*$ *for all* $x, y \in \mathcal{A}$.

(d) $\|x^*x\| = \|x\|^2$ *for all* $x \in \mathcal{A}$.

Any mapping $x \mapsto x^*$ on an algebra satisfying (a), (b), and (c) is called an involution on the algebra. The element x^* is usually called the adjoint of x.

We point out that condition (d) above is the most stringent condition in the definition of a C^*-algebra.

The involution in a C^*-algebra is an isometry. In fact, for every x in $\mathcal{A}$ we have

$$\|x\|^2 = \|x^*x\| \le \|x^*\|\,\|x\|,$$

so that $\|x\| \le \|x^*\|$. Replacing x by x^* we conclude that $\|x^*\| \le \|x\|$ and hence $\|x\| = \|x^*\|$.

It can also be shown that any involution $x \mapsto x^*$ satisfying $\|x\|^2 \le \|x^*x\|$ must be an isometry. In fact, this condition gives

$$\|x\|^2 \le \|x^*x\| \le \|x^*\|\,\|x\|$$

and hence $\|x\| \le \|x^*\|$. This implies that $\|x\| = \|x^*\|$ for all x in $\mathcal{A}$.

Let $\mathbf{1}$ be the (multiplicative) unit in a C^*-algebra $\mathcal{A}$. For every x in $\mathcal{A}$ we have

$$\mathbf{1}^*x = (x^*\mathbf{1})^* = x, \quad x\mathbf{1}^* = (\mathbf{1}x^*)^* = x.$$

By the uniqueness of the unit in an algebra, $\mathbf{1}^* = \mathbf{1}$.

It is easy to check that an element x in a C^*-algebra is invertible if and only if x^* is. In this case we also have $(x^*)^{-1} = (x^{-1})^*$. This clearly implies that

$$\sigma(x^*) = \overline{\sigma(x)} = \{\bar{\lambda} : \lambda \in \sigma(x)\},$$

and $r(x^*) = r(x)$.

8.2 Examples

The following are several examples of C^*-algebras.

EXAMPLE 8.2
$\mathcal{A} = \mathbf{C}$ with $z^* = \bar{z}$ is the simplest C^*-algebra. □

EXAMPLE 8.3
$\mathcal{A} = \mathbf{C}(K)$ with $f^* = \bar{f}$ is a commutative C^*-algebra. We shall show later that every commutative C^*-algebra is of this form. Note that if K consists of n distinct points, then $\mathbf{C}(K) = \mathbf{C}^n$. Thus there exists a commutative C^*-algebra of any given dimension. □

EXAMPLE 8.4
$\mathcal{A} = L^\infty$ with $f^* = \bar{f}$ is a commutative C^*-algebra. □

EXAMPLE 8.5
$\mathcal{A} = B(H)$ with the usual adjoint operation as involution is a C^*-algebra. If H is n-dimensional, then $B(H) = M_n(\mathbf{C})$ is the algebra of all $n \times n$ complex matrices and the adjoint of a matrix in $M_n(\mathbf{C})$ is its conjugate transpose. □

EXAMPLE 8.6
Suppose $\mathcal{A}$ is a C^*-algebra and $\mathcal{B}$ is a closed (unital) subalgebra of $\mathcal{A}$ that is closed under the involution. Then $\mathcal{B}$ is a C^*-algebra by itself with the norm, involution, and algebraic structure inherited from $\mathcal{A}$. We call such $\mathcal{B}$ a C^*-subalgebra of $\mathcal{A}$. We shall show later that every C^*-algebra is a C^*-subalgebra of $B(H)$. It is also true that every finite dimensional C^*-algebra is a C^*-subalgebra of some $M_n(\mathbf{C})$. See [11]. □

8.3 Special Elements in a C^*-Algebra

DEFINITION 8.7 *Suppose $\mathcal{A}$ is a C^*-algebra and x is in $\mathcal{A}$.*

(a) *We say that x is self-adjoint if $x^* = x$.*

(b) *We say that x is unitary if $x^*x = xx^* = \mathbf{1}$, or equivalently $x^* = x^{-1}$.*

(c) *We say that x is normal if $x^*x = xx^*$.*

(d) *We say that x is positive if $x = y^*y$ for some y in $\mathcal{A}$.*

(e) *We say that x is a projection if $x^* = x = x^2$.*

It is clear that both self-adjoint and unitary elements are normal. It is also clear that positive elements are self-adjoint and projections are positive. We write $x \geq 0$ if and only if x is positive. Every unitary element is a unit vector.

A subset of a C^*-algebra will be said to be self-adjoint if it is closed under the involution, that is, if it contains all the adjoints of its elements.

THEOREM 8.1
If x is normal in a C^*-algebra $\mathcal{A}$, then $r(x) = \|x\|$.

PROOF First assume that x is self-adjoint. In this case,

$$\|x^2\| = \|x^*x\| = \|x\|^2$$

and so

$$\|x^{2^n}\| = \|x\|^{2^n}, \qquad n \geq 1.$$

By the spectral radius formula (Theorem 5.5),

$$r(x) = \lim_{n\to\infty} \|x^{2^n}\|^{\frac{1}{2^n}} = \|x\|.$$

To prove the formula above for x normal, we observe that if x and y are elements in a Banach algebra with $xy = yx$, then the spectral radius formula easily implies that $r(xy) \leq r(x)r(y)$. Now if x is normal in a C^*-algebra, then x commutes with x^*. By the observation above and the self-adjoint case

$$\|x\|^2 = \|x^*x\| = r(x^*x) \leq r(x^*)r(x).$$

Since $r(x^*) = r(x)$ and $r(x) \leq \|x\|$, we conclude that $r(x) = \|x\|$ for all normal elements x in $\mathcal{A}$. ∎

As a consequence of the theorem above we see that for every normal element x in a C^*-algebra the spectrum $\sigma(x)$ contains at least one point on the circle $|z| = \|x\|$. It follows that if x is normal and $x^n = 0$ for some positive integer

n, then $x = 0$. This is because $x^n = 0$ implies that $\sigma(x) = \{0\}$ (by the spectral mapping theorem in **5.3**) and hence $\|x\| = r(x) = 0$.

Another consequence of the formula $r(x) = \|x\|$ for x normal is that the norm in a C^*-algebra is unique. Namely, given an involution $x \mapsto x^*$ on an algebra $\mathcal{A}$ there is at most one norm making $\mathcal{A}$ a C^*-algebra. In fact, if $\| \ \|$ makes $\mathcal{A}$ a C^*-algebra, then

$$\|x\|^2 = \|x^*x\| = r(x^*x) = \sup\{|\lambda| : \lambda \in \sigma(x^*x)\}.$$

Thus the norm is uniquely determined by the algebraic structure of $\mathcal{A}$ and by the particular involution used.

Suppose x and y are elements in a C^*-algebra $\mathcal{A}$. We say that x is unitarily equivalent to y if $x = uyu^*$ for some unitary element u in $\mathcal{A}$. It is easy to see that unitary equivalence is indeed an equivalence relation. It is also clear that unitarily equivalent elements have the same norm and spectrum. In fact, unitary equivalence preserves all C^*-algebraic properties of an element.

PROPOSITION 8.2

Suppose x is in a C^*-algebra $\mathcal{A}$.

(a) *If x is self-adjoint, then $\sigma(x) \subset \mathbf{R}$.*

(b) *If x is unitary, then $\sigma(x) \subset \partial\mathbf{D}$.*

PROOF First assume that x is unitary. Then $\|x\| = 1$ and

$$\sigma(x) \subset \{\lambda : |\lambda| \leq 1\}.$$

It follows that

$$\sigma(x^{-1}) = \{\lambda^{-1} : \lambda \in \sigma(x)\} \subset \{\lambda : |\lambda| \geq 1\}.$$

But $x^{-1} = x^*$ and $\sigma(x^*) = \overline{\sigma(x)}$, we conclude that

$$\sigma(x) \subset \{\lambda : |\lambda| \geq 1\}$$

and hence $\sigma(x) \subset \partial\mathbf{D}$.

Next suppose x is self-adjoint. Let $y = \exp(ix)$. By power series multiplication we have $yy^* = y^*y = \mathbf{1}$ and so y is unitary. By (b) and the spectral mapping theorem, $\exp(i\lambda)$ is unimodulus for every λ in $\sigma(x)$. This clearly implies that $\sigma(x)$ is contained in the real line $\mathbf{R}$. ∎

REMARK 8.3 It is possible to prove (a) in the proposition above without the use of (b) and the spectral mapping theorem. For this, suppose x is self-adjoint and $c = a + ib \in \sigma(x)$. For each integer n, let

$$x_n = x - a\mathbf{1} + inb\mathbf{1}$$

and observe that (the definition of the spectrum is sufficient)

$$i(n+1)b = a + ib - a + inb \in \sigma(x_n).$$

Accordingly,

$$\begin{aligned}(n^2+2n+1)b^2 &= \left|i(n+1)b\right|^2 \le \left(r(x_n)\right)^2 \le \|x_n\|^2 \\ &= \|x_n^* x_n\| = \left\|(x-a\mathbf{1})^2 + n^2b^2\mathbf{1}\right\| \\ &\le \|x-a\mathbf{1}\|^2 + n^2b^2.\end{aligned}$$

Thus

$$(2n+1)b^2 \le \|x-a\mathbf{1}\|^2, \qquad n = 1, 2, \cdots.$$

This clearly implies $b = 0$, and so $c = a \in \mathbf{R}$. ∎

Exercises

8.1 If x is self-adjoint and $y = \exp(ix)$, then y is unitary.

8.2 Every x in a C^*-algebra can be uniquely written as $x = x_1 + ix_2$ with x_1 and x_2 self-adjoint. Furthermore, x is normal if and only if $x_1x_2 = x_2x_1$.

8.3 Given a Banach algebra $\mathcal{A}$ how many involutions (if any) are there on $\mathcal{A}$ making $\mathcal{A}$ a C^*-algebra?

8.4 If x is a projection in a C^*-algebra and if $x \neq 0, \mathbf{1}$, then $\sigma(x) = \{0, 1\}$.

8.5 Suppose $\mathcal{A}$ is a C^*-algebra and $x \in \mathcal{A}$ satisfies $x^* = -x$. Show that $\sigma(x)$ is contained in the imaginary axis of the complex plane.

8.6 Show that $x = 0$ if and only if $xx^* = 0$ if and only if $x^*x = 0$ in a C^*-algebra.

8.7 If x is positive in a C^*-algebra $\mathcal{A}$, then yxy^* is positive for every y in $\mathcal{A}$.

8.8 Show that a matrix A in $M_n(\mathbf{C})$ is diagonalizable (by a unitary) if and only if A is normal.

8.9 Suppose $\mathcal{A} = B(H)$ and $T \in \mathcal{A}$. Show that T is normal if and only if $\|Tx\| = \|T^*x\|$ for every $x \in H$; T is unitary if and only if T is a surjective isometry; and T is a projection if and only if T is the identity on the orthogonal compliment of its kernel.

9

Commutative C^*-Algebras

9.1 C^*-Homomorphisms

DEFINITION 9.1 *Suppose $\mathcal{A}$ and $\mathcal{B}$ are C^*-algebras. A mapping $\Phi : \mathcal{A} \to \mathcal{B}$ is said to be a C^*-homomorphism if*

(a) $\Phi(ax + by) = a\Phi(x) + b\Phi(y)$ *for all* $a, b \in \mathbf{C}$ *and* $x, y \in \mathcal{A}$.

(b) $\Phi(xy) = \Phi(x)\Phi(y)$ *for all* $x, y \in \mathcal{A}$.

(c) $\Phi(x^*) = \Phi(x)^*$ *for all* $x \in \mathcal{A}$.

(d) Φ *maps the unit in* $\mathcal{A}$ *to the unit in* $\mathcal{B}$.

If Φ is further one-to-one, we say that Φ is a C^*-isomorphism. Two C^*-algebras are C^*-isomorphic if there exists a C^*-isomorphism from one onto the other.

The following result shows that C^*-homomorphisms are always continuous.

PROPOSITION 9.1
Suppose $\mathcal{A}$ and $\mathcal{B}$ are C^*-algebras and $\Phi : \mathcal{A} \to \mathcal{B}$ is a C^*-homomorphism. For every $x \in \mathcal{A}$ we have $\sigma\big(\Phi(x)\big) \subset \sigma(x)$ and $\big\|\Phi(x)\big\| \leq \|x\|$.

PROOF The containment $\sigma\big(\Phi(x)\big) \subset \sigma(x)$ is clear. This containment, together with the spectral radius formula for normal elements (Theorem 8.1), also gives

$$\begin{aligned}
\|x\|^2 &= \|x^*x\| = r(x^*x) \\
&\geq r\big(\Phi(x^*x)\big) = r\big(\Phi(x)^*\Phi(x)\big) \\
&= \big\|\Phi(x)^*\Phi(x)\big\| = \big\|\Phi(x)\big\|^2,
\end{aligned}$$

completing the proof of the proposition. ∎

COROLLARY 9.2
Every C^*-homomorphism is contractive and hence bounded.

COROLLARY 9.3
Suppose Φ is a C^*-isomorphism from $\mathcal{A}$ onto $\mathcal{B}$. Then $\sigma\big(\Phi(x)\big) = \sigma(x)$ and $\big\|\Phi(x)\big\| = \|x\|$ for all $x \in \mathcal{A}$.

PROOF This is because Φ^{-1} is also a C^*-homomorphism. ∎

We shall see later that this last corollary also holds when Φ is only a C^*-isomorphism from $\mathcal{A}$ *into* $\mathcal{B}$.

9.2 Representation of Commutative C^*-Algebras

THEOREM 9.4
Every commutative C^*-algebra is C^*-isomorphic to $\mathbf{C}(K)$ for some compact Hausdorff space K. Specifically, for a commutative C^*-algebra $\mathcal{A}$ the Gelfand transform is a C^*-isomorphism from $\mathcal{A}$ onto $\mathbf{C}(\mathcal{M}_{\mathcal{A}})$.

PROOF Let $\Gamma : \mathcal{A} \to \mathbf{C}(\mathcal{M}_{\mathcal{A}})$ be the Gelfand transform. We know that Γ is an algebraic homomorphism. We need to show Γ is one-to-one, onto, and preserves the involutions.

For each x in $\mathcal{A}$ let

$$x_1 = \frac{x + x^*}{2}, \qquad x_2 = \frac{x - x^*}{2i}.$$

Clearly x_1 and x_2 are self-adjoint, $x = x_1 + ix_2$, and $x^* = x_1 - ix_2$. Since $\mathcal{A}$ is commutative, for $k = 0$ and 1 the range of $\Gamma(x_k)$ is equal to $\sigma(x_k)$ (see Corollary 5.2), which is contained in $\mathbf{R}$ by Proposition 8.2. It follows that

$$\Gamma(x^*) = \Gamma(x_1) - i\Gamma(x_2) = \overline{\Gamma(x_1) + i\Gamma(x_2)} = \overline{\Gamma(x)},$$

so that Γ preserves the involutions.

That Γ is one-to-one follows from Corollary 5.2 and the spectral radius formula for normal elements (Theorem 8.1). In fact,

$$\begin{aligned} \|\Gamma(x)\|_\infty^2 &= \|\Gamma(x^*)\Gamma(x)\|_\infty = \|\Gamma(x^*x)\|_\infty \\ &= r(x^*x) = \|x^*x\| = \|x\|^2. \end{aligned}$$

So Γ is an isometry. In particular, Γ is one-to-one and the image of Γ in $\mathbf{C}(\mathcal{M}_{\mathcal{A}})$ is closed.

Finally, it is easy to check that the image of Γ in $\mathbf{C}(\mathcal{M}_{\mathcal{A}})$ is a closed and self-adjoint subalgebra of $\mathbf{C}(\mathcal{M}_{\mathcal{A}})$ which contains the constant functions and separates points in $\mathcal{M}_{\mathcal{A}}$. By the Stone-Weierstrass approximation theorem, Γ is onto. ∎

COROLLARY 9.5
Suppose x is in a commutative C^*-algebra $\mathcal{A}$.

(a) *x is self-adjoint if and only if $\sigma(x) \subset \mathbf{R}$.*

(b) *x is unitary if and only if $\sigma(x) \subset \partial\mathbf{D}$.*

(c) *x is positive if and only if $\sigma(x) \subset \mathbf{R}^+$.*

(d) *x is a projection if and only if $\sigma(x) \subset \{0, 1\}$.*

9.3 An Example

Recall that every L^∞ space is a commutative C^* algebra. By Theorem 9.4 it is C^*-isomorphic to $\mathbf{C}(K)$ for some compact Hausdorff space K. We show that such spaces K have very interesting topological properties.

A topological space X will be called "extremely disconnected" if every open set in X has an open closure.

THEOREM 9.6
Let $\mathcal{A}$ be an L^∞ space. Then $\mathcal{A}$ is C^*-isomorphic to $\mathbf{C}(K)$, where K is an extremely disconnected compact Hausdorff space.

PROOF Let K be the maximal ideal space of $\mathcal{A}$. Then $\mathcal{A}$ is C^*-isomorphic to $\mathbf{C}(K)$ because $\mathcal{A}$ is commutative. We already know that K is compact Hausdorff. To show that K is extremely disconnected let U be an open set in K and consider

$$\mathcal{F} = \{f \in \mathbf{C}(K) : 0 \le f \le 1, f \equiv 0 \text{ on } K - U\}.$$

Let $\Gamma : \mathcal{A} \to \mathbf{C}(K)$ be the Gelfand transform. Since $\mathcal{A}$ is an L^∞ space and Γ^{-1} preserves the order structure, $\Gamma^{-1}(\mathcal{F})$ is a bounded set in $\mathcal{A}$ and hence has a least upper bound. It follows that $\mathcal{F}$ has a least upper bound f_0 in $\mathbf{C}(K)$. It is clear that $0 \le f_0 \le 1$. For every x in U there exists (by Uryson's lemma) a function f in $\mathcal{F}$ such that $f(x) = 1$, so that $f_0 \equiv 1$ on U. By continuity, $f_0 \equiv 1$ on $\overline{U}$, the closure of U. On the other hand, if x is in $K - \overline{U}$, then by Uryson's lemma again there exists a function g in $\mathbf{C}(K)$ such that $0 \le g \le 1$, $g(x) = 0$, and $g \equiv 1$ on $\overline{U}$, so that g is an upper bound for $\mathcal{F}$, $f_0 \le g$, and so $f_0(x) = 0$. It follows that $f_0 \equiv 0$ on $K - \overline{U}$ and $f_0 \equiv 1$ on $\overline{U}$. Since f_0 is continuous, we conclude that the closure of U is open. ∎

9.4 Subalgebras and the Spectrum

In the remainder of this lecture we consider the dependence of the spectrum on the containing algebra. We prove the result in the general setting of Banach algebras, although we will only need the C^*-algebra version.

LEMMA 9.7
Suppose U and V are open sets in $\mathbf{C}$, $U \subset V$, and V contains no boundary points of U. Then U is the union of certain connected components of V.

PROOF This is clear. ▮

LEMMA 9.8
Suppose $\mathcal{A}$ is a Banach algebra and $\{x_n\}$ is a sequence of invertible elements in $\mathcal{A}$ which converges to a noninvertible element x. Then $\|x_n^{-1}\| \to +\infty$.

PROOF Assume the contrary. Then we may as well assume that $\|x_n^{-1}\| \leq C$ for some constant $C > 0$ and all $n \geq 1$. Since

$$\|\mathbf{1} - x_n^{-1}x\| = \|x_n^{-1}(x_n - x)\| \leq C\|x_n - x\|,$$

we have $\|\mathbf{1} - x_n^{-1}x\| < 1$ for large enough n and hence $x_n^{-1}x$ is invertible for large enough n, contradicting the assumption that x is not invertible. ▮

THEOREM 9.9
Suppose $\mathcal{A}$ is a closed subalgebra of a Banach algebra $\mathcal{B}$ with $\mathbf{1} \in \mathcal{A}$. If x is in $\mathcal{A}$, then $\sigma_{\mathcal{A}}(x)$ is the union of $\sigma_{\mathcal{B}}(x)$ and some (possibly none) bounded components of the complement of $\sigma_{\mathcal{B}}(x)$ in $\mathbf{C}$.

PROOF Let $\Omega_{\mathcal{A}}$ and $\Omega_{\mathcal{B}}$ be the complements of $\sigma_{\mathcal{A}}(x)$ and $\sigma_{\mathcal{B}}(x)$ in $\mathbf{C}$, respectively. Since $\sigma_{\mathcal{B}}(x) \subset \sigma_{\mathcal{A}}(x)$, we have $\Omega_{\mathcal{A}} \subset \Omega_{\mathcal{B}}$.

Let λ_0 be a boundary point of $\Omega_{\mathcal{A}}$. There exists a sequence $\{\lambda_n\}$ in $\Omega_{\mathcal{A}}$ such that $\lambda_n \to \lambda_0$. Note that $x_n = \lambda_n\mathbf{1} - x$ is invertible in $\mathcal{A}$, while the limit $\lambda_0\mathbf{1} - x$ is not invertible in $\mathcal{A}$. By Lemma 9.8,

$$\|(\lambda_n\mathbf{1} - x)^{-1}\| \to \infty.$$

If λ_0 is in $\Omega_{\mathcal{B}}$, $\lambda_0\mathbf{1} - x$ is invertible in $\mathcal{B}$, and the continuity of the inversion implies

$$(\lambda_n\mathbf{1} - x)^{-1} \to (\lambda_0\mathbf{1} - x)^{-1},$$

a contradiction to the fact that

$$\|(\lambda_n\mathbf{1} - x)^{-1}\| \to +\infty.$$

We conclude that $\Omega_{\mathcal{B}}$ does not contain any boundary points of $\Omega_{\mathcal{A}}$. According to Lemma 9.7, $\Omega_{\mathcal{A}}$ is the union of certain connected components of $\Omega_{\mathcal{B}}$, so that $\sigma_{\mathcal{A}}(x)$ is the union of $\sigma_{\mathcal{B}}(x)$ and some bounded components of the complement of $\sigma_{\mathcal{B}}(x)$ in **C**. ∎

THEOREM 9.10

Suppose $\mathcal{A}$ is a (unital) C^*-subalgebra of a C^*-algebra $\mathcal{B}$ and x is in $\mathcal{A}$. Then x is invertible in $\mathcal{A}$ if and only if x is invertible in $\mathcal{B}$.

PROOF If x is invertible in $\mathcal{A}$, clearly x is invertible in $\mathcal{B}$. On the other hand, if x is invertible in $\mathcal{B}$, then so is $y = x^*x$, so that $\sigma_{\mathcal{B}}(y)$ does not contain 0. Since y is self-adjoint in $\mathcal{B}$, $\sigma_{\mathcal{B}}(y) \subset \mathbf{R}$, and so the complement of $\sigma_{\mathcal{B}}(y)$ has no bounded components. By Theorem 9.9 we have $\sigma_{\mathcal{B}}(y) = \sigma_{\mathcal{A}}(y)$. Thus $\sigma_{\mathcal{A}}(y)$ does not contain 0, and hence y is invertible in $\mathcal{A}$. It follows that $x^{-1} = (x^*x)^{-1}x^*$ belongs to $\mathcal{A}$. ∎

COROLLARY 9.11

Suppose $\mathcal{A}$ is a (unital) C^*-subalgebra of a C^*-algebra $\mathcal{B}$ and x is in $\mathcal{A}$. Then $\sigma_{\mathcal{A}}(x) = \sigma_{\mathcal{B}}(x)$.

Exercises

9.1 Show that a C^*-algebra $\mathcal{A}$ is commutative if and only if its Gelfand transform is isometric and preserves the involutions.

9.2 Construct a Banach algebra $\mathcal{B}$ and a unital subalgebra $\mathcal{A}$ such that $\sigma_{\mathcal{A}}(x) \neq \sigma_{\mathcal{B}}(x)$ for some x in $\mathcal{A}$.

9.3 A projection in a C^*-algebra is said to be nontrivial if it is different from 0 and **1**. Show that a commutative C^*-algebra contains nontrivial projections if and only if its maximal ideal space is disconnected.

9.4 Let $\mathcal{A} = \mathbf{C}(\overline{\mathbf{D}})$, where $\overline{\mathbf{D}}$ is the closed unit disc in **C**. Show that the quotient $\mathbf{C}(\overline{\mathbf{D}})/\mathbf{C}_0(\mathbf{D})$ is a C^*-algebra. Identify its maximal ideal space.

9.5 Let K be a compact Hausdorff space. Show that every closed ideal of $\mathbf{C}(K)$ is of the form

$$I = \{f \in \mathbf{C}(K) : f \equiv 0 \text{ on } F\},$$

where F is a closed subset of K. When is such an ideal maximal?

9.6 Is every algebraic homomorphism from a Banach algebra into another necessarily bounded?

9.7 Suppose Φ is an algebraic homomorphism from a C^*-algebra into another C^*-algebra. If $\|\Phi(x)\| \le \|x\|$ for all x, then Φ is a C^*-homomorphism.

10

The Spectral Theorem and Applications

10.1 The Spectral Theorem

Suppose $\mathcal{A}$ is a C^*-algebra and S is a subset of $\mathcal{A}$. The C^*-algebra generated by S, denoted by $\mathcal{A}[S]$, is the smallest C^*-subalgebra of $\mathcal{A}$ containing S. If $S = \{x_1, \cdots, x_n\}$, we write $\mathcal{A}[S] = \mathcal{A}[x_1, \cdots, x_n]$. In particular, $\mathcal{A}[x]$ is the C^*-subalgebra generated by x.

PROPOSITION 10.1

If x is normal in a C^*-algebra $\mathcal{A}$, then $\mathcal{A}[x]$ is commutative. Moreover, an element $y \in \mathcal{A}$ belongs to $\mathcal{A}[x]$ if and only if y can be approximated in norm by polynomials in x and x^*.

PROOF This is clear. ∎

THEOREM 10.2 THE SPECTRAL THEOREM

Suppose x is normal in a C^*-algebra $\mathcal{A}$. The maximal ideal space of $\mathcal{A}[x]$ is homeomophic to $\sigma(x)$. Moreover, if we identify the maximal ideal space of $\mathcal{A}[x]$ with $\sigma(x)$, the Gelfand transform Γ on $\mathcal{A}[x]$ has the property that $\Gamma\big(p(x, x^*)\big) = p(z, \bar{z})$ for every polynomial p of two variables.

PROOF Let $\mathcal{M}$ be the maximal ideal space of $\mathcal{A}[x]$ and let

$$\Gamma : \mathcal{A}[x] \to \mathbf{C}(\mathcal{M})$$

be the Gelfand transform. Since $\mathcal{A}[x]$ is commutative, Γ is a surjective C^*-isomorphism (see Theorem 9.4). Define

$$\Phi : \mathcal{M} \to \sigma(x)$$

by

$$\Phi(\varphi) = \Gamma(x)(\varphi) = \varphi(x), \qquad \varphi \in \mathcal{M}.$$

By Corollaries 5.2 and 9.11, Φ is well-defined and onto.

Suppose $\Phi(\varphi_1) = \Phi(\varphi_2)$ or $\varphi_1(x) = \varphi_2(x)$. Since

$$\varphi_k(x^*) = \Gamma(x^*)(\varphi_k) = \overline{\Gamma(x)(\varphi_k)} = \overline{\varphi_k(x)}, \quad k = 1, 2,$$

we have $\varphi_1(x^*) = \varphi_2(x^*)$. This clearly implies that

$$\varphi_1\big(p(x, x^*)\big) = \varphi_2\big(p(x, x^*)\big)$$

for every polynomial p of two variables. Recall that such polynomials are dense in $\mathcal{A}[x]$, we see that $\varphi_1 = \varphi_2$ and hence Φ is one-to-one.

If $\varphi_\alpha \to \varphi$ in $\mathcal{M}$, then $\varphi_\alpha(y) \to \varphi(y)$ for every y in $\mathcal{A}[x]$. In particular, $\varphi_\alpha(x) \to \varphi(x)$. Thus $\Phi(\varphi_\alpha) \to \Phi(\varphi)$ and Φ is continuous. Being a bijective continuous function from a compact Hausdorff space to another compact Hausdorff space, Φ must be a homeomorphism.

If we identify the maximal ideal space $\mathcal{M}$ of $\mathcal{A}[x]$ with $\sigma(x)$, the Gelfand transform

$$\Gamma : \mathcal{A}[x] \to \mathbf{C}\big(\sigma(x)\big)$$

satisfies (from the definition of Φ) $\Gamma(x)(z) = z$ for all $z \in \sigma(x)$. Since Γ is a C^*-homomorphism,

$$\Gamma\big(p(x, x^*)\big)(z) = p(z, \bar{z})$$

for every polynomial p of two variables. ∎

10.2 An Example

Let $H = L^2(\mathbf{D}, dA)$, where $\mathbf{D}$ is the open unit disk and dA is the normalized area measure on $\mathbf{D}$. Let $\mathcal{A} = B(H)$, the C^*-algebra of all bounded linear operators on $L^2(\mathbf{D}, dA)$. Suppose f is a continuous function on the closed unit disk $\overline{\mathbf{D}}$ and $x = M_f$ is the multiplication operator on $H = L^2(\mathbf{D}, dA)$ induced by f. It is easy to check that x is normal in $\mathcal{A}$ and x^* is simply $M_{\bar{f}}$. Thus $\mathcal{A}[x]$ is the closure of the set of multiplication operators induced by functions of the form $p(f, \bar{f})$, where p is a polynormial of two variables. The norm of the multiplication operator induced by $p(f, \bar{f})$ coincides with

$$\sup\Big\{\big|p(f, \bar{f})(z)\big| : z \in \mathbf{D}\Big\} = \sup\big\{|p(z, \bar{z})| : z \in K\big\},$$

where $K = f(\overline{\mathbf{D}})$ is the spectrum of x. From this and the Stone-Weierstrass approximation theorem we conclude that the mapping

$$M_{p(f,\bar{f})} \leftrightarrow p(z, \bar{z})$$

extends to a C^*-isomorphism Φ from $\mathcal{A}[x]$ onto $\mathbf{C}(K)$. Moreover, if we identify the maximal ideal space of $\mathcal{A}[x]$ with K, then Φ is simply the Gelfand transform.

It is easy to generalize the above idea to the case where f is just bounded and measurable on $\mathbf{D}$. The same idea can also be generalized to other situations involving multiplication operators on L^2 induced by bounded functions. In fact, it is known that every normal operator is unitarily equivalent to a multiplication operator on some L^2 space of the complex plane; see [1].

10.3 The Continuous Functional Calculus

Suppose x is normal in a C^*-algebra $\mathcal{A}$. We always identify the maximal ideal space of $\mathcal{A}[x]$ with $\sigma(x)$. Under this identification the Gelfand transform

$$\Gamma : \mathcal{A}[x] \to \mathbf{C}(\sigma(x))$$

is a surjective C^*-isomorphism. This allows us to define $f(x)$ for every function f in $\mathbf{C}(\sigma(x))$.

DEFINITION 10.1 *Suppose x is normal in a C^*-algebra $\mathcal{A}$. For every f in $\mathbf{C}(\sigma(x))$ we define $f(x) = \Gamma^{-1}(f)$. The mapping $f \mapsto f(x)$ from $\mathbf{C}(\sigma(x))$ onto $\mathcal{A}[x]$ is called the continuous functional calculus for x.*

The following theorem lists some of the basic properties of the continuous functional calculus.

THEOREM 10.3
Suppose x is normal in a C^*-algebra $\mathcal{A}$. The continuous functional calculus for x has the following properties:

(a) *$f \mapsto f(x)$ is a C^*-isomorphism from $\mathbf{C}(\sigma(x))$ onto $\mathcal{A}[x]$.*

(b) *If $f(z) = p(z, \bar{z})$ with p a polynomial of two variables, then $f(x) = p(x, x^*)$. In particular, if $f(z) = z$ is the identity function on $\sigma(x)$, we have $f(x) = x$.*

(c) *$\sigma(f(x)) = f(\sigma(x))$ for all f in $\mathbf{C}(\sigma(x))$.*

(d) *If Φ is a C^*-homomorphism from $\mathcal{A}$ into another C^*-algebra, then*

$$\Phi(f(x)) = f(\Phi(x))$$

for all f in $\mathbf{C}(\sigma(x))$.

PROOF These properties are clear from the definition of the continuous functional calculus. ∎

10.4 Some Applications

Next we derive some consequences of the spectral theorem and the continuous functional calculus.

THEOREM 10.4
Suppose x is normal in a C^*-algebra $\mathcal{A}$.

(a) *x is self-adjoint if and only if $\sigma(x) \subset \mathbf{R}$.*

(b) *x is unitary if and only if $\sigma(x) \subset \partial\mathbf{D}$.*

(c) *x is a projection if and only if $\sigma(x) \subset \{0, 1\}$.*

PROOF This is clear from the spectral theorem and the continuous functional calculus for x. ∎

In the next lecture we will also show that if x is normal in a C^*-algebra $\mathcal{A}$ then x is positive if and only if $\sigma(x) \subset \mathbf{R}^+$.

THEOREM 10.5
Suppose x is positive in a C^*-algebra $\mathcal{A}$. There exists a unique positive y in $\mathcal{A}$ such that $y^2 = x$.

PROOF Let $f(t) = \sqrt{t}$. Since x is self-adjoint and $\sigma(x) \subset \mathbf{R}^+$, f is in $\mathbf{C}\big(\sigma(x)\big)$, and the continuous functional calculus implies that $y = f(x) \geq 0$ with $y^2 = x$.

To see that such y is unique, assume that $x = \widetilde{y}^2$ for some other positive $\widetilde{y}$ in $\mathcal{A}$. Let $\{p_n\}$ be a sequence of polynomials converging uniformly to $f(t) = \sqrt{t}$ on $\sigma(x)$. Let $q_n(t) = p_n(t^2)$. Since $\sigma(\widetilde{y}) \subset \mathbf{R}^+$ and

$$\sigma(x) = \sigma(\widetilde{y}^2) = \sigma(\widetilde{y})^2 = \big\{t^2 : t \in \sigma(\widetilde{y})\big\},$$

we have

$$\lim_{n\to\infty} q_n(t) = \lim_{n\to\infty} p_n(t^2) = f(t^2) = t$$

uniformly for t in $\sigma(\widetilde{y})$. By the continuous functional calculus for $\widetilde{y}$

$$\widetilde{y} = \lim_{n\to\infty} q_n(\widetilde{y}) = \lim_{n\to\infty} p_n(\widetilde{y}^2) = \lim_{n\to\infty} p_n(x) = f(x) = y,$$

proving the uniqueness of the square root. ∎

The positive element y in the theorem above is called *the* positive square root of x and will be denoted by $\sqrt{x}$ or $x^{\frac{1}{2}}$. For every element x in a C^*-algebra we write $|x| = \sqrt{x^*x}$. It is clear that $|x| \geq 0$ for all x.

THEOREM 10.6

Every element x in a C^*-algebra $\mathcal{A}$ is a finite linear combination of unitary elements.

PROOF We know that $x = x_1 + ix_2$ with x_1 and x_2 self-adjoint. So it suffices to prove the result for x self-adjoint. It can further be assumed that $\|x\| \leq 1$. In this case, $\sigma(x) \subset [-1, 1]$ and the function

$$f(t) = t + i\sqrt{1 - t^2}$$

belongs to $\mathbf{C}\big(\sigma(x)\big)$. Let $u = f(x)$ via the continuous functional calculus. Since

$$f(t)\overline{f(t)} = 1, \quad t = \frac{f(t) + \overline{f(t)}}{2},$$

for every $t \in [-1, 1]$, the continuous functional calculus implies that u is unitary and $x = (u + u^*)/2$. ∎

THEOREM 10.7

Suppose $\Phi : \mathcal{A} \to \mathcal{B}$ is an injective C^*-homomorphism from a C^*-algebra $\mathcal{A}$ into a C^*-algebra $\mathcal{B}$. Then $\big\|\Phi(x)\big\| = \|x\|$ and $\sigma\big(\Phi(x)\big) = \sigma(x)$ for every x in $\mathcal{A}$.

PROOF We first assume x is self-adjoint. It is clear that $\sigma\big(\Phi(x)\big) \subset \sigma(x)$. If strict inclusion occurs, there exists a nonzero function f in $\mathbf{C}\big(\sigma(x)\big)$ such that $f \equiv 0$ on $\sigma\big(\Phi(x)\big)$. By the continuous functional calculus,

$$f(x) \neq 0, \qquad \Phi\big(f(x)\big) = f\big(\Phi(x)\big) = 0,$$

a contradiction to the fact that Φ is injective. This shows that $\sigma\big(\Phi(x)\big) = \sigma(x)$ and hence $r(x) = r\big(\Phi(x)\big)$. By the spectral radius formula for normal elements (Theorem 8.1), we have $\big\|\Phi(x)\big\| = \|x\|$.

For general x we have

$$\|x\|^2 = \|x^*x\| = \big\|\Phi(x^*x)\big\| = \big\|\Phi(x)^*\Phi(x)\big\| = \big\|\Phi(x)\big\|^2.$$

Thus Φ is isometric and its image must be closed. The proof of the theorem is complete in view of Corollary 9.11. ∎

As a consequence of the theorem above we see that the image of an injective C^*-homomorphism is always closed. This is actually true for any C^*-homomorphism, as will be shown in the next lecture.

Exercises

10.1 If x is normal in a C^*-algebra $\mathcal{A}$, then x is a projection if and only if $x = x^2$.

10.2 If x is unitary in a C^*-algebra $\mathcal{A}$ and $\sigma(x) \neq \partial\mathbf{D}$, then $x = e^{iy}$ for some self-adjoint y in $\mathcal{A}$.

10.3 If x is invertible, there exists $\epsilon > 0$ such that $\epsilon\mathbf{1} \leq x^*x \leq \epsilon^{-1}\mathbf{1}$.

10.4 If x is normal and $\sigma(x) \subset \mathbf{R}^+$, then $x \geq 0$.

10.5 Show that if x is self-adjoint, then $|x| \pm x \geq 0$, and $\|x\|\mathbf{1} \pm x \geq 0$.

10.6 If $0 \leq x \leq y$, then $\sqrt{x} \leq \sqrt{y}$. If $0 \leq x \leq y$ and x is invertible, then y is invertible with $y^{-1} \leq x^{-1}$.

10.7 Suppose x is normal and $\sigma(x)$ is contained in the imaginary axis of the complex plane. Show that $x^* = -x$. The inverse is also true.

10.8 Let x be an element in a C^*-algebra $\mathcal{A}$. Then x, x^*, and $\mathbf{1}$ are linearly dependent if and only if x is normal and $\sigma(x)$ is contained in a straightline in $\mathbf{C}$.

10.9 Suppose T is a normal operator on a Hilbert space H. Show that there exists a self-adjoint operator S on H such that $T = f(S)$, where f is a continuous function from the spectrum of S into $\mathbf{C}$.

11

Further Applications

11.1 The Image of a C^*-Homomorphism

In this lecture we give several more applications of the spectral theorem and the continuous functional calculus for normal elements in a C^*-algebra.

THEOREM 11.1
Suppose $\mathcal{A}$ and $\mathcal{B}$ are C^*-algebras and $\Phi : \mathcal{A} \to \mathcal{B}$ is a C^*-homomorphism. Then the image of Φ is closed in $\mathcal{B}$.

PROOF Suppose y is in $\mathcal{B}$ and $\|\Phi(x_n) - y\| \to 0$ for some sequence $\{x_n\}$ in $\mathcal{A}$. We need to show that $y = \Phi(x)$ for some x in $\mathcal{A}$. By expressing y and x_n in terms of their "real and imaginary" parts, we may as well assume that y and x_n are all self-adjoint. Passing to a subsequence if necessary, we may also assume that

$$\left\|\Phi(x_{n+1}) - \Phi(x_n)\right\| < \frac{1}{2^n}, \qquad n \geq 1.$$

For each $n \geq 1$ let

$$f_n(t) = \begin{cases} \frac{1}{2^n}, & t \geq \frac{1}{2^n} \\ t, & -\frac{1}{2^n} < t < \frac{1}{2^n} \\ -\frac{1}{2^n}, & t \leq -\frac{1}{2^n}. \end{cases}$$

Since $\Phi(x_{n+1}) - \Phi(x_n)$ is self-adjoint and has norm less than 2^{-n}, f_n is the identity function on $\sigma\big(\Phi(x_{n+1}) - \Phi(x_n)\big)$. By the continuous functional calculus,

$$\begin{aligned} \Phi(x_{n+1}) - \Phi(x_n) &= f_n\big(\Phi(x_{n+1}) - \Phi(x_n)\big) \\ &= f_n\big(\Phi(x_{n+1} - x_n)\big) \\ &= \Phi\big(f_n(x_{n+1} - x_n)\big). \end{aligned}$$

Also by the continuous functional calculus

$$\|f_n(x_{n+1}-x_n)\| = \sup\{|f(t)| : t \in \sigma(x_{n+1}-x_n)\} \le \frac{1}{2^n}.$$

Let

$$x = x_1 + \sum_{n=1}^{\infty} f_n(x_{n+1}-x_n).$$

Then x belongs to $\mathcal{A}$ and

$$\begin{aligned}\Phi(x) &= \Phi(x_1) + \sum_{n=1}^{\infty} \Phi(f_n(x_{n+1}-x_n)) \\ &= \Phi(x_1) + \sum_{n=1}^{\infty} f_n\big(\Phi(x_{n+1}) - \Phi(x_n)\big) \\ &= \Phi(x_1) + \sum_{n=1}^{\infty} \big(\Phi(x_{n+1}) - \Phi(x_n)\big) \\ &= \lim_{n\to\infty} \Phi(x_n),\end{aligned}$$

completing the proof of the theorem. ∎

11.2 Positive and Negative Parts of an Element

THEOREM 11.2

Suppose x is self-adjoint in a C^*-algebra $\mathcal{A}$. There exist unique self-adjoint elements x^+ and x^- in $\mathcal{A}$ such that

(a) $x = x^+ - x^-$.

(b) $\sigma(x^+), \sigma(x^-) \subset \mathbf{R}^+$. *(We shall see later that this is equivalent to the statement that x^+ and x^- are positive.)*

(c) $x^+x^- = x^-x^+ = 0$.

(d) $\|x\| = \max\big(\|x^+\|, \|x^-\|\big)$.

PROOF Let $f(t) = t$, $t \in \mathbf{R}$, and let

$$f^+(t) = \begin{cases} t, & t \ge 0 \\ 0, & t \le 0, \end{cases} \qquad f^-(t) = \begin{cases} 0, & t \ge 0 \\ -t, & t \le 0. \end{cases}$$

It is clear that

$$f(t) = f^+(t) - f^-(t), \qquad t \in \mathbf{R},$$

and both f^+ and f^- are nonnegative. By the continuous functional calculus

$$f(x) = f^+(x) - f^-(x) \quad \text{or} \quad x = x^+ - x^-$$

with

$$x^+ = f^+(x) = \left(\sqrt{f^+}(x)\right)^2 \geq 0, \quad x^- = f^-(x) = \left(\sqrt{f^-}(x)\right)^2 \geq 0,$$

and

$$\sigma(x^+) = \{f^+(t) : t \in \sigma(x)\} \subset \mathbf{R}^+, \ \sigma(x^-) = \{f^-(t) : t \in \sigma(x)\} \subset \mathbf{R}^+.$$

Also

$$x^+x^- = f^+(x)f^-(x) = (f^+f^-)(x) = 0,$$

and

$$\begin{aligned} \|x\| &= \sup\{|f(t)| : t \in \sigma(x)\} \\ &= \sup\{\max\left(f^+(t), f^-(t)\right) : t \in \sigma(x)\} \\ &= \max\left(\|x^+\|, \|x^-\|\right). \end{aligned}$$

To prove the uniqueness of x^+ and x^-, suppose that $x = x_1 - x_2$, where x_1 and x_2 are self-adjoint in $\mathcal{A}$ with nonnegative spectra and $x_1x_2 = x_2x_1 = 0$. It is then clear that

$$x^n = x_1^n + (-x_2)^n$$

for all positive integers n and hence

$$p(x) = p(x_1) + p(-x_2)$$

for every polynomial p without constant term. Since $f^+(t) = 0$, the Stone-Weierstrass approximation theorem implies that there is a sequence $\{p_n\}$ of such polynomials that converges to f^+ uniformly on the union of $\sigma(x)$, $\sigma(x_1)$, and $\sigma(-x_2)$. Thus

$$f^+(x) = \lim p_n(x) = \lim\left[p_n(x_1) + p(-x_2)\right] = f^+(x_1) + f^+(-x_2).$$

Since $f^+(t) \equiv t$ on $\sigma(x_1)$ and $f^+(t) \equiv 0$ on $\sigma(-x_2)$, we have $f^+(x_1) = x_1$ and $f^+(-x_2) = 0$. It follows that $x_1 = f^+(x) = x^+$ and $x_2 = x^-$. ∎

11.3 Positivity in Terms of the Spectrum

Finally in this lecture we show that a normal element in a C^*-algebra is positive if and only if its spectrum is contained in $\mathbf{R}^+$. We need the following lemma which is of some independent interest.

LEMMA 11.3

Suppose x is self-adjoint in a C^*-algebra $\mathcal{A}$ and $a \geq \|x\|$. Then $\sigma(x) \subset \mathbf{R}^+$ if and only if $\|x - a\mathbf{1}\| \leq a$.

PROOF Since x is self-adjoint and $\|x\| \leq a$, the spectrum of x is contained in $[-a, a]$. By the spectral radius formula for normal elements (Theorem 8.1)

$$\|x - a\mathbf{1}\| = r(x - a\mathbf{1}) = \sup\{a - \lambda : \lambda \in \sigma(x)\}.$$

We see that $\|x - a\mathbf{1}\| \leq a$ if and only if $\sigma(x) \subset \mathbf{R}^+$. ∎

COROLLARY 11.4

If x, y are self-adjoint elements in a C^*-algebra $\mathcal{A}$ and $\sigma(x), \sigma(y) \subset \mathbf{R}^+$, then $\sigma(x + y) \subset \mathbf{R}^+$.

PROOF By Lemma 11.3 and the containments $\sigma(x), \sigma(y) \subset \mathbf{R}^+$ we have

$$\big\|x - \|x\|\mathbf{1}\big\| \leq \|x\|, \quad \big\|y - \|y\|\mathbf{1}\big\| \leq \|y\|.$$

By the triangle inequality

$$\big\|(x + y) - \big(\|x\| + \|y\|\big)\mathbf{1}\big\| \leq \|x\| + \|y\|.$$

Since

$$\|x + y\| \leq \|x\| + \|y\|,$$

Lemma 11.3 again implies that $\sigma(x + y) \subset \mathbf{R}^+$. ∎

THEOREM 11.5

Suppose x is normal in a C^*-algebra $\mathcal{A}$. Then x is positive if and only if $\sigma(x) \subset \mathbf{R}^+$.

PROOF If $\sigma(x) \subset \mathbf{R}^+$, the function $f(t) = t$ is continuous and nonnegative on $\sigma(x)$ and hence $\sqrt{f}$ belongs to $\mathbf{C}\big(\sigma(x)\big)$. By the continuous functional calculus, $x = y^2 = y^*y$ with $y = (\sqrt{f})(x)$, so that x is positive.

If x is positive, there exists y in $\mathcal{A}$ such that $x = y^*y$. Clearly x is self-adjoint. By Theorem 11.2 we can write $x = x^+ - x^-$ with $x^+, x^- \geq 0$, $x^+x^- = x^-x^+ = 0$, and $\sigma(x^+), \sigma(x^-) \subset \mathbf{R}^+$. Let $v = yx^-$. Since $x^+x^- = 0$, we have

$$v^*v = x^-xx^- = x^-(x^+ - x^-)x^- = -(x^-)^3.$$

This implies that $\sigma(v^*v) \subset \mathbf{R}^-$. Applying Corollary 3.5, we conclude that $\sigma(vv^*) \subset \mathbf{R}^-$.

On the other hand, we can write $v = v_1 + iv_2$ with v_1 and v_2 self-adjoint. An easy calculation shows that

$$v^*v + vv^* = 2(v_1^2 + v_2^2),$$

or

$$v^*v = 2(v_1^2 + v_2^2) - vv^*.$$

The right-hand side of the identity above is a sum of three self-adjoint elements whose spectra are all contained in $\mathbf{R}^+$. By Corollary 11.4, $\sigma(v^*v) \subset \mathbf{R}^+$. Combining this with the inclusion $\sigma(v^*v) \subset \mathbf{R}^-$ established in the previous paragraph, we obtain $\sigma(v^*v) = \{0\}$, so that

$$\|v^*v\| = r(v^*v) = 0.$$

It follows that $(x^-)^3 = 0$ and hence $x^- = 0$ by the spectral theorem. Therefore, $x = x^+$ and $\sigma(x) = \sigma(x^+) \subset \mathbf{R}^+$. ∎

Exercises

11.1 Every normal element in a C^*-algebra can be written as a linear combination of four commuting positive elements.

11.2 If x is self-adjoint, show that $|x| = x^+ + x^-$.

11.3 Suppose Φ is a C^*-homomorphism from a C^*-algebra $\mathcal{A}$ into another C^*-algebra $\mathcal{B}$. Show that

$$\|\Phi(x)\| = \inf\{\|x - y\|_{\mathcal{A}} : \Phi(y) = 0\}$$

for every x in $\mathcal{A}$.

11.4 Show that the set of positive elements in a C^*-algebra is a cone, called the positive cone of $\mathcal{A}$. That is, if $x_1, x_2 \geq 0$ in $\mathcal{A}$ and $a \geq 0$ in $\mathbf{R}$, then $x_1 + x_2 \geq 0$ and $ax_1 \geq 0$.

12

Polar Decomposition

12.1 Operators on Hilbert Spaces

In this lecture we concentrate on the special C^*-algebra $B(H)$, where H is a Hilbert space. The unit in the algebra $B(H)$ will be denoted by I. Our main results will be a new characterization for positive operators in $B(H)$ and the polar decomposition for operators in $B(H)$. We point out that polar decomposition cannot be carried out in an arbitary C^*-algebra .

LEMMA 12.1
An operator T in $B(H)$ is self-adjoint if and only if $\langle Tx, x\rangle$ is real for all x in H.

PROOF This follows from a standard polarization argument. We leave the proof as an exercise. ∎

LEMMA 12.2
For every T in $B(H)$ we have

$$\ker T = (\operatorname{Ran} T^*)^\perp, \quad \overline{\operatorname{Ran} T} = (\ker T^*)^\perp,$$

where $\ker T$ *is the kernel of T and* $\operatorname{Ran} T$ *is the range of T.*

PROOF This is again an easy and standard result in functional analysis. We leave it as an exercise. ∎

LEMMA 12.3
An operator T in $B(H)$ is invertible if and only if T has dense range and $\|Tx\| \geq \epsilon\|x\|$ for some constant $\epsilon > 0$ and all $x \in H$.

PROOF If T is invertible (so that T^{-1} is also bounded), the range of T is the whole space H and

$$\|Tx\| \geq \|T^{-1}\|^{-1}\|x\|$$

for all x in H.

If $\|Tx\| \geq \epsilon\|x\|$ for some $\epsilon > 0$ and all $x \in H$ and if the range of T is dense, we show that T is invertible. Since $\|Tx\| \geq \epsilon\|x\|$ implies that T is one-to-one, we need only show that the range of T is closed. Suppose $Tx_n \to y$ in H. The inequality

$$\|x_n - x_m\| \leq \frac{1}{\epsilon}\|Tx_n - Tx_m\|$$

shows that $\{x_n\}$ is a Cauchy sequence. Since H is complete, there exists an x in H such that $x_n \to x$. By the continuity of T

$$y = \lim_{n\to\infty} Tx_n = Tx,$$

so that $\operatorname{Ran} T$ is closed. ∎

An operator T in $B(H)$ satisfying $\|Tx\| \geq \epsilon\|x\|$ for some $\epsilon > 0$ and all $x \in H$ is said to be bounded below. The following result is clearly a consequence of the previous two lemmas.

COROLLARY 12.4

Suppose T is in $B(H)$ and both T and T^* are bounded below. Then T is invertible.

THEOREM 12.5

Suppose T is in the C^*-algebra $B(H)$. Then T is positive if and only if $\langle Tx, x\rangle \geq 0$ for all x in H.

PROOF If T is positive, then there exists S in $B(H)$ such that $T = S^*S$, so that

$$\langle Tx, x\rangle = \|Sx\|^2 \geq 0$$

for all x in H.

If $\langle Tx, x\rangle \geq 0$ for all x in H, T is self-adjoint by Lemma 12.1, so that $\sigma(T) \subset \mathbf{R}$. For each $\lambda < 0$ and $x \in H$ we have

$$\left\|(T - \lambda I)x\right\|^2 = \|Tx\|^2 - 2\lambda\langle Tx, x\rangle + \lambda^2\|x\|^2 \geq \lambda^2\|x\|^2.$$

Since $T - \lambda I$ is self-adjoint, Lemma 12.4 implies that $T - \lambda I$ is invertible. This shows that $\sigma(T) \subset \mathbf{R}^+$, and hence T is positive. ∎

12.2 Partial Isometries

DEFINITION 12.1 *Suppose T is an operator in $B(H)$. If $\|Tx\| = \|x\|$ for every x in H, T is called an isometry. If $\|Tx\| = \|x\|$ for all $x \perp \ker T$, we say that T is a partial isometry. If T is a partial isometry, the space $(\ker T)^{\perp}$ is called the initial space of T and the range of T is called the final space of T.*

It is clear that all isometries and projections are partial isometries. The range of a partial isometry is necessarily closed. Every partial isometry is necessarily a contraction, namely, $\|Tx\| \leq \|x\|$ for all x in H. If H is finite dimensional, every isometry is onto and hence a unitary. This is no longer true when the dimension of H is infinite.

The following are several examples of partial isometries.

EXAMPLE 12.2

Let H be a separable infinite dimensional Hilbert space. Let $\{e_n\}$ be an orthonormal basis for H. Define an operator T on H as follows:

$$T\left(\sum_{n=1}^{\infty} a_n e_n\right) = \sum_{n=1}^{\infty} a_n e_{n+1}, \qquad \{a_n\} \in l^2.$$

Then T is an isometry which is not onto. This operator is called the forward shift. □

EXAMPLE 12.3

Again let H be a separable infinite dimensional Hilbert space with an orthonormal basis $\{e_n\}$. Let T be the operator on H defined by

$$T\left(\sum_{n=1}^{\infty} a_n e_n\right) = \sum_{n=1}^{\infty} a_{n+1} e_n, \qquad \{a_n\} \in l^2.$$

Then T is a partial isometry with initial space being the orthogonal complement of the vector e_1. This operator is called the backward shift. □

EXAMPLE 12.4

Suppose X_1 and X_2 are two closed subspaces of a Hilbert space H with the same dimension. Then there exists a partial isometry with initial space X_1 and range X_2. In fact, if $\{e_\alpha\}_{\alpha\in A}$ and $\{\sigma_\alpha\}_{\alpha\in A}$ are orthonormal bases for X_1 and X_2, respectively, we can define a bounded linear operator T on H as follows:

$$Tx = 0,\ x \perp X_1, \quad \text{and} \quad T\left(\sum_{\alpha\in A} a_\alpha e_\alpha\right) = \sum_{\alpha\in A} a_\alpha \sigma_\alpha,\ \sum_{\alpha\in A} |a_\alpha|^2 < +\infty.$$

It is then clear that T is a partial isometry with initial space X_1 and final space X_2. □

PROPOSITION 12.6

An operator T in $B(H)$ is a partial isometry if and only if T^*T is a projection (onto the initial space of T).

PROOF First assume that T is a partial isometry. The inequality

$$\big\langle (I - T^*T)x, x \big\rangle = \|x\|^2 - \|Tx\|^2 \geq 0, \quad x \in H,$$

shows that $I - T^*T \geq 0$. If $x \perp \ker T$, then $\|Tx\| = \|x\|$ and hence

$$\big\|\sqrt{I - T^*T}x\big\|^2 = \big\langle (I - T^*T)x, x \big\rangle = 0.$$

Since

$$\big\|(I - T^*T)x\big\| \leq \big\|\sqrt{I - T^*T}\big\| \, \big\|\sqrt{I - T^*T}x\big\|,$$

we see that $(I - T^*T)x = 0$ or $T^*Tx = x$. Thus T^*T is the orthogonal projection onto $(\ker T)^{\perp}$, the initial space of T.

Conversely, if T^*T is a projection onto X, then

$$\|Tx\|^2 = \langle T^*Tx, x\rangle = \begin{cases} \|x\|^2, & x \in X \\ 0, & x \perp X. \end{cases}$$

So T is a partial isometry with initial space X. ∎

COROLLARY 12.7

An operator T in $B(H)$ is a partial isometry if and only if T^* is a partial isometry.

PROOF Suppose T is a partial isometry. By Proposition 12.6 $P = T^*T$ is a projection onto the initial space of T, so that $T(T^*T) = T$. Thus

$$(TT^*)^2 = T(T^*T)T^* = TT^*$$

and hence TT^* is a projection. Applying Proposition 12.6 again we see that T^* is a partial isometry. ∎

12.3 Polar Decomposition

Recall that every complex number z can be written as $z = r\zeta$, where $r \geq 0$ and ζ is unimodulus. This is usually called the polar representation for complex numbers. The following theorem is called the polar decomposition for bounded operators on Hilbert spaces.

THEOREM 12.8

For every operator T in $B(H)$ there exists a positive operator P and a partial isometry V such that $T = VP$. Moreover, V and P are unique if $\ker P = \ker V$.

PROOF Let $P = |T| = \sqrt{T^*T}$. We have

$$\|Px\|^2 = \langle P^*Px, x\rangle = \langle T^*Tx, x\rangle = \|Tx\|, \quad x \in H.$$

Define $\widetilde{V}$ on $\operatorname{Ran} P$ by $\widetilde{V}(Px) = Tx$, $x \in H$. Then $\widetilde{V}$ extends to an isometry V from the closure of $\operatorname{Ran} P$ into H. If we also define $Vx = 0$ for x orthogonal to the closure of $\operatorname{Ran} P$, V becomes a partial isometry with initial space being the closure of $\operatorname{Ran} P$. For each x in H we have $Tx = \widetilde{V}Px = VPx$, so that $T = VP$ with

$$\ker V = (\operatorname{Ran} P)^{\perp} = \ker P.$$

To prove uniqueness, assume that we also have a polar decomposition $T = WQ$, where $Q \geq 0$ and W is a partial isometry with $\ker W = \ker Q$. Since W^*W is a projection onto

$$(\ker W)^{\perp} = (\ker Q)^{\perp} = \overline{\operatorname{Ran} Q}$$

by Proposition 12.6, we have

$$P^2 = T^*T = QW^*WQ = Q^2.$$

Thus $P = Q$ by the uniqueness of the positive square root. It follows that $WP = VP$ and so $W = V$ on $(\ker P)^{\perp} = \overline{\operatorname{Ran} P}$. Since

$$(\operatorname{Ran} P)^{\perp} = \ker P = \ker W = \ker V,$$

we also have $W = V = 0$ on $(\operatorname{Ran} P)^{\perp}$. This shows that $W = V$ on H and the polar decomposition of T is unique. ∎

COROLLARY 12.9

For every operator T in $B(H)$ there exists a positive operator Q and a partial isometry W such that $T = QW$. Moreover, the operators Q and W are unique if $\operatorname{Ran} W = \overline{\operatorname{Ran} Q}$.

PROOF By Theorem 12.8 there exists a partial isometry V and a positive operator P such that $T^* = VP$. Taking adjoints we obtain $T = PV^*$, which is the form we desire with $W = V^*$ and $Q = P$ (see Corollary 12.7). Since

$$\operatorname{Ran} W = \overline{\operatorname{Ran} Q} = [\ker Q]^{\perp}$$

if and only if

$$\ker V = \ker W^* = [\operatorname{Ran} W]^\perp = [\ker Q]^{\perp\perp} = \ker P,$$

the uniqueness follows from Theorem 12.8 again. ∎

Exercises

12.1 If x is invertible in a C^*-algebra $\mathcal{A}$, there exist positive x_1 and unitary x_2 in $\mathcal{A}$ such that $x = x_1 x_2$. The decomposition is unique. Use this decomposition to show that the invertible group of the matrix algebra $M_n(\mathbf{C})$ is path connected.

12.2 Construct a C^*-algebra $\mathcal{A}$ such that there exists $x \in \mathcal{A}$ which cannot be decomposed into $x = x_1 x_2$ with x_1 positive and $x_2^* x_2$ being a projection.

12.3 Prove Lemmas 12.1 and 12.2.

12.4 When is the sum of two partial isometries again a partial isometry?

12.5 Show that an operator T on H is an isometry if and only if $T^*T = I$. If T is an isometry on H, then T is a unitary if and only if V^* is an isometry if and only if $\ker T^* = \{0\}$.

13

Positive Linear Functionals and States

13.1 Definitions and Basic Properties

DEFINITION 13.1 *Suppose $\mathcal{A}$ is a C^*-algebra and φ is a linear functional on $\mathcal{A}$.*

(a) *We say that φ is Hermitian if $\varphi(x^*) = \overline{\varphi(x)}$ for all $x \in \mathcal{A}$.*

(b) *We say that φ is positive if $\varphi(x) \geq 0$ for all $x \geq 0$.*

(c) *We say that φ is a state if φ is positive and $\varphi(\mathbf{1}) = 1$.*

It can be shown that Hermitian functionals are always bounded (see Exercise 13.7). Also we shall see a little later that positive functionals are Hermitian.

PROPOSITION 13.1

A linear functional φ on a C^*-algebra $\mathcal{A}$ is Hermitian if and only if $\varphi(x)$ is real for every self-adjoint x.

PROOF It is clear that a Hermitian linear functional φ on $\mathcal{A}$ satisfies $\varphi(x) \in \mathbf{R}$ for all self-adjoint x in $\mathcal{A}$. On the other hand, if $\varphi(x) \in \mathbf{R}$ for every $x^* = x$, we show that φ is Hermitian. Given x in $\mathcal{A}$ we can write $x = x_1 + ix_2$ with both x_1 and x_2 self-adjoint. It follows that

$$\varphi(x^*) = \varphi(x_1) - i\varphi(x_2) = \overline{\varphi(x)},$$

so that φ is Hermitian. ∎

PROPOSITION 13.2

All positive linear functionals on a C^*-algebra are Hermitian.

PROOF Suppose φ is a positive linear functional on a C^*-algebra $\mathcal{A}$ and x is self-adjoint in $\mathcal{A}$. We have

$$\varphi(|x|+x) \geq 0, \qquad \varphi(|x|-x) \geq 0,$$

because $|x|+x \geq 0$ and $|x|-x \geq 0$ by the continuous functional calculus. It follows that

$$\varphi(x) = \frac{1}{2}\left(\varphi(|x|+x) - \varphi(|x|-x)\right)$$

is real, so that φ is Hermitian by the previous proposition. ∎

PROPOSITION 13.3
If φ is a Hermitian linear functional on a C^*-algebra $\mathcal{A}$, then

$$\|\varphi\| = \sup\{\varphi(x) : x^* = x, \|x\| \leq 1\}.$$

PROOF Let

$$M = \sup\{\varphi(x) : x^* = x, \|x\| \leq 1\}.$$

It is clear that $M \leq \|\varphi\|$. Given $\epsilon > 0$ there exists a unit vector x in $\mathcal{A}$ such that $\|\varphi\| - \epsilon < |\varphi(x)|$. Multiplying x by a unimodulus scalar if necessary, we may assume that $\varphi(x) \geq 0$. In this case,

$$\|\varphi\| - \epsilon < \varphi(x) = \overline{\varphi(x)} = \varphi(x^*).$$

Write $x = x_1 + ix_2$, where

$$x_1 = \frac{x+x^*}{2}, \qquad x_2 = \frac{x-x^*}{2i}$$

are self-adjoint. The identity $\varphi(x) = \varphi(x^*)$ and the fact that φ is Hermitian easily imply that $\varphi(x_2) = 0$. It follows that $\|\varphi\| - \epsilon < \varphi(x_1)$ with

$$\|x_1\| = \left\|\frac{x+x^*}{2}\right\| \leq \|x\| = 1.$$

Thus $\|\varphi\| - \epsilon < M$, and therefore $\|\varphi\| \leq M$. Combining this with $M \leq \|\varphi\|$, we conclude that $\|\varphi\| = M$. ∎

PROPOSITION 13.4
For every positive linear functional φ on a C^*-algebra $\mathcal{A}$ we have

$$\left|\varphi(x^*y)\right|^2 \leq \varphi(x^*x)\varphi(y^*y), \quad x, y \in \mathcal{A}.$$

PROOF Since φ is a positive linear functional on $\mathcal{A}$,

$$\langle x, y\rangle = \varphi(y^*x), \qquad x, y \in \mathcal{A}$$

defines a semi-inner product on the linear space $\mathcal{A}$. The desired result is then a consequence of the Cauchy-Schwarz inequality. ∎

13.2 A Characterization of Positivity

THEOREM 13.5

Suppose φ is a linear functional on a C^*-algebra $\mathcal{A}$. Then φ is positive if and only if φ is bounded with $\|\varphi\| = \varphi(\mathbf{1})$.

PROOF First assume that φ is positive. In particular, it is Hermitian. Let x be a self-adjoint unit vector in $\mathcal{A}$. Clearly $\mathbf{1} - x \geq 0$. This implies that $\varphi(\mathbf{1} - x) \geq 0$ or $\varphi(x) \leq \varphi(\mathbf{1})$. By Proposition 13.3, φ is bounded with $\|\varphi\| \leq \varphi(\mathbf{1})$. But

$$\varphi(\mathbf{1}) \leq \|\varphi\|\|\mathbf{1}\| = \|\varphi\|.$$

We must have $\|\varphi\| = \varphi(\mathbf{1})$.

Next assume that φ is bounded and $\|\varphi\| = \varphi(\mathbf{1})$. We proceed to show that φ is positive. Without loss of generality we may also assume $\|\varphi\| = \varphi(\mathbf{1}) = 1$.

Given $x \geq 0$ in $\mathcal{A}$. Write $\varphi(x) = a + ib$ with a and b both real. We need to show that $a \geq 0$ and $b = 0$. Choose $\epsilon > 0$ small enough so that $\sigma(\mathbf{1}-\epsilon x) \subset [0, 1]$. This is possible because

$$\sigma(\mathbf{1} - \epsilon x) = \{1 - \epsilon\lambda : \lambda \in \sigma(x)\}$$

and $\sigma(x) \subset \mathbf{R}^+$ is compact. By the spectral radius formula for normal elements (Theorem 8.1) we have

$$\|\mathbf{1} - \epsilon x\| = r(\mathbf{1} - \epsilon x) \leq 1,$$

and

$$1 - \epsilon a \leq |1 - \epsilon(a + ib)| = |\varphi(\mathbf{1} - \epsilon x)| \leq 1.$$

This clearly shows that $a \geq 0$.

To see $b = 0$ we consider the sequence

$$x_n = x - a\mathbf{1} + inb\mathbf{1}, \qquad n \geq 1.$$

Note that

$$\varphi(x_n) = (n + 1)ib$$

and so

$$(n + 1)^2 b^2 = |\varphi(x_n)|^2 \leq \|x_n\|^2.$$

On the other hand,

$$\|x_n\|^2 = \|x_n^* x_n\| = \left\|(x - a\mathbf{1})^2 + n^2 b^2 \mathbf{1}\right\| \leq \|x - a\mathbf{1}\|^2 + n^2 b^2.$$

It follows that

$$(n+1)^2 b^2 \leq \|x - a\mathbf{1}\|^2 + n^2 b^2$$

or

$$(2n+1) b^2 \leq \|x - a\mathbf{1}\|^2$$

for all n, which clearly shows $b = 0$. ∎

COROLLARY 13.6
A linear functional φ on a C^*-algebra $\mathcal{A}$ is a state if and only if $\|\varphi\| = \varphi(\mathbf{1}) = 1$.

13.3 The State Space

For every C^*-algebra $\mathcal{A}$ we let $S(\mathcal{A})$ denote the space of all states on $\mathcal{A}$. Note that $S(\mathcal{A})$ is a subset of the closed unit ball in the dual space of $\mathcal{A}$. We topologize $S(\mathcal{A})$ with the weak-star topology inherited from the dual space of $\mathcal{A}$. The space $S(\mathcal{A})$ together with the weak-star topology will be called the state space of $\mathcal{A}$. We first show that the state space $S(\mathcal{A})$ is always nonempty.

THEOREM 13.7
Suppose $\mathcal{A}$ is a C^*-algebra and x is in $\mathcal{A}$. For each $\lambda \in \sigma(x)$ there exists a state φ on $\mathcal{A}$ such that $\varphi(x) = \lambda$.

PROOF Fix $x \in \mathcal{A}$ and $\lambda \in \sigma(x)$. Consider

$$M = \{ax + b\mathbf{1} : a, b \in \mathbf{C}\}.$$

M is a subspace of $\mathcal{A}$. Define $\varphi_0 : M \to \mathbf{C}$ by

$$\varphi_0(ax + b\mathbf{1}) = a\lambda + b, \qquad a, b \in \mathbf{C}.$$

It is easy to see that φ_0 is a linear functional on M with $\varphi_0(x) = \lambda$, $\varphi_0(\mathbf{1}) = 1$, and

$$\sup\left\{|\varphi_0(ax + b\mathbf{1})| : \|ax + b\mathbf{1}\| \leq 1\right\} = 1.$$

In fact, for $a, b \in \mathbf{C}$ we have

$$a\lambda + b \in \sigma(ax + b\mathbf{1})$$

and so

$$|a\lambda + b| \leq r(ax + b\mathbf{1}) \leq \|ax + b\mathbf{1}\|;$$

equality here is achieved by taking $a = 0$ and $b = 1$. By the Hahn-Banach extension theorem, φ_0 extends to a bounded linear functional φ on $\mathcal{A}$ with $\|\varphi\| = 1$. The linear functional φ satisfies $\varphi(x) = \lambda$ and is a state in view of Corollary 13.6. ∎

PROPOSITION 13.8

Suppose $\mathcal{A}$ is a C^*-algebra. Then the state space $S(\mathcal{A})$ of $\mathcal{A}$ is a convex, compact, and Hausdorff space.

PROOF Let $\varphi_1, \varphi_2 \in S(\mathcal{A})$, $t \in (0,1)$, and $\varphi = t\varphi_1 + (1-t)\varphi_2$. Clearly φ is again linear and $\|\varphi\| \leq 1$. Also $\varphi(\mathbf{1}) = 1$, which implies that $\|\varphi\| \geq 1$. It follows that $\|\varphi\| = \varphi(\mathbf{1}) = 1$. This shows that φ is again in $S(\mathcal{A})$ and hence $S(\mathcal{A})$ is convex.

That $S(\mathcal{A})$ is Hausdorff is clear.

Finally, since

$$S(\mathcal{A}) = \{\varphi \in (\mathcal{A}^*)_1 : \varphi(x) \geq 0, x \geq 0, \varphi(\mathbf{1}) = 1\}$$

is closed in the compact space $(\mathcal{A}^*)_1$, $S(\mathcal{A})$ must be compact in the weak-star topology. ∎

The following result shows that the state space of a C^*-algebra is not only nonempty, it is also large enough to reveal many properties of an element in the algebra.

THEOREM 13.9

Suppose $\mathcal{A}$ is a C^*-algebra and x is in $\mathcal{A}$.

(a) *$x = 0$ if and only if $\varphi(x) = 0$ for all $\varphi \in S(\mathcal{A})$.*

(b) *x is self-adjoint if and only if $\varphi(x) \in \mathbf{R}$ for all $\varphi \in S(\mathcal{A})$.*

(c) *x is positive if and only if $\varphi(x) \geq 0$ for all $\varphi \in S(\mathcal{A})$.*

(d) *If x is normal, then $\|x\| = |\varphi(x)|$ for some $\varphi \in S(\mathcal{A})$.*

PROOF (a) Suppose $\varphi(x) = 0$ for every $\varphi \in S(\mathcal{A})$. Then $\varphi(x_k) = 0$ for $k = 1, 2$ and $\varphi \in S(\mathcal{A})$, where $x = x_1 + ix_2$ with x_1 and x_2 self-adjoint. By Theorem 13.7, $\sigma(x_k) = \{0\}$ for $k = 1, 2$. This implies that $x_1 = x_2 = 0$ and so $x = 0$.

(b) If $\varphi(x) \in \mathbf{R}$ for all $\varphi \in S(\mathcal{A})$, then $\varphi(x - x^*) = 0$ for all $\varphi \in S(\mathcal{A})$. Thus $x = x^*$ by (a), or x is self-adjoint.

(c) Assume $\varphi(x) \geq 0$ for all $\varphi \in S(\mathcal{A})$. Then x is self-adjoint by (b) and $\sigma(x) \subset \mathbf{R}^+$ by Theorem 13.7. It follows from the spectral theorem that $x \geq 0$.

(d) Finally if x is normal, then $\|x\| = r(x)$ and the compactness of $\sigma(x)$ implies that there exists $\lambda \in \sigma(x)$ such that $\|x\| = |\lambda|$. By Theorem 13.7 again, there exists $\varphi \in S(\mathcal{A})$ such that $\|x\| = |\varphi(x)|$. ∎

13.4 Pure States

Since the state space $S(\mathcal{A})$ is convex and weak-star compact, it has extreme points. In fact, by the Krein-Milman theorem, $S(\mathcal{A})$ is the weak-star closed convex hull of the set $P(\mathcal{A})$ of its extreme points. Elements of $P(\mathcal{A})$ will be called pure states of $\mathcal{A}$. Thus every φ in $S(\mathcal{A})$ can be approximated in the weak-star topology by elements of the form $t_1\varphi_1 + \cdots + t_n\varphi_n$, where each φ_k is in $P(\mathcal{A})$, each t_k is in $(0, 1)$, and $t_1 + \cdots + t_n = 1$.

LEMMA 13.10
Suppose S is a nonempty compact convex set in a locally convex space X and φ is a continuous linear functional on X. Then there exists an extreme point x_0 of S such that $\operatorname{Re}\varphi(x) \le \operatorname{Re}\varphi(x_0)$ *for all $x \in S$.*

PROOF Let $f(x) = \operatorname{Re}\varphi(x)$, $x \in X$. Then f is a real-valued continuous function on the compact set S. Let S_0 be the set of points in S where f achieves the maximum value on S. Then S_0 is a nonempty compact convex subset of S, so that S_0 has extreme points, by the Krein-Milman theorem. It is easy to check that each extreme point of S_0 is also an extreme point of S and so has the desired properties. ∎

The following result shows that the set of pure states of a C^*-algebra is also sufficiently large.

THEOREM 13.11
Suppose $\mathcal{A}$ is a C^*-algebra and $x \in \mathcal{A}$.

(a) *$x = 0$ if and only if $\varphi(x) = 0$ for every $\varphi \in P(\mathcal{A})$.*

(b) *x is self-adjoint if and only if $\varphi(x) \in \mathbf{R}$ for every $\varphi \in P(\mathcal{A})$.*

(c) *x is positive if and only if $\varphi(x) \ge 0$ for every $\varphi \in P(\mathcal{A})$.*

(d) *If x is normal, then there exists $\varphi \in P(\mathcal{A})$ such that $\|x\| = |\varphi(x)|$.*

PROOF The statements (a), (b), and (c) all follow from Theorem 13.9 and the Krein-Milman theorem. We proceed to prove (d).

Suppose x is normal and, by Theorem 13.9, suppose φ_0 is a state of $\mathcal{A}$ such that $\|x\| = \lambda\varphi_0(x)$, where λ is a unimodulus complex number. Let Φ be the weak-star continuous linear functional on $\mathcal{A}^*$ defined by $\Phi(\varphi) = \lambda\varphi(x)$, $\varphi \in \mathcal{A}^*$. By Lemma 13.10 there exists $\varphi_1 \in P(\mathcal{A})$ such that $\operatorname{Re}\Phi(\varphi) \leq \operatorname{Re}\Phi(\varphi_1)$ for all $\varphi \in S(\mathcal{A})$. In particular,

$$\|x\| = \operatorname{Re}\Phi(\varphi_0) \leq \operatorname{Re}\Phi(\varphi_1) \leq \big|\varphi_1(x)\big| \leq \|x\|,$$

so that $\|x\| = \big|\varphi_1(x)\big|$ with $\varphi_1 \in P(\mathcal{A})$. ∎

Exercises

13.1 Suppose H is a Hilbert space. For each unit vector x in H the functional $T \mapsto \langle Tx, x\rangle$ is a state of $B(H)$. Such states are called vector states.

13.2 If φ is a state of $\mathcal{A}$ and u is a unitary in $\mathcal{A}$, then $x \mapsto \varphi(uxu^*)$ is a state of $\mathcal{A}$.

13.3 Show that every multiplicative linear functional on a C^*-algebra is a pure state.

13.4 If $\mathcal{A}$ is a commutative C^*-algebra, then $P(\mathcal{A})$ consists exactly of the multiplicative linear functionals.

13.5 Show that if x is normal in a C^*-algebra, then

$$\sigma(x) = \big\{\varphi(x) : \varphi \in P(\mathcal{A})\big\}.$$

13.6 Let x be a self-adjoint element in a C^*-algebra $\mathcal{A}$ and let φ be a state of $\mathcal{A}$ such that $\varphi(x^2) = \varphi(x)^2$. Show that $\varphi(xy) = \varphi(x)\varphi(y)$ for all y in $\mathcal{A}$.

13.7 Show that every Hermitian linear functional on a C^*-algebra is bounded.

14

The GNS Construction

14.1 The Construction

In this lecture we show that every C^*-algebra is C^*-isomorphic to a C^*-subalgebra of $B(H)$ for some Hilbert space H. The proof will be constructive. This construction is usually called the GNS (standing for Gelfand, Neumark, and Segal) construction.

PROPOSITION 14.1
Suppose φ is a state on a C^*-algebra $\mathcal{A}$. Let

$$\mathcal{L}_\varphi = \{x \in \mathcal{A} : \varphi(x^*x) = 0\}.$$

Then $\mathcal{L}_\varphi$ is a closed left ideal in $\mathcal{A}$. Moreover, $\varphi(x^*y) = 0$ whenever x or y is in $\mathcal{L}_\varphi$.

PROOF Recall that

$$|\varphi(x^*y)|^2 \leq \varphi(x^*x)\varphi(y^*y), \qquad x,\, y \in \mathcal{A}.$$

Thus $\varphi(x^*y) = 0$ if either x or y belongs to $\mathcal{L}_\varphi$. It follows from this and the identity

$$\varphi\big((ax+by)^*(ax+by)\big) = |a|^2\varphi(x^*x) + 2\mathrm{Re}\,\big(\bar{a}b\varphi(x^*y)\big) + |b|^2\varphi(y^*y)$$

that $\mathcal{L}_\varphi$ is a linear subspace of $\mathcal{A}$. $\mathcal{L}_\varphi$ is closed because the function $x \mapsto \varphi(x^*x)$ is continuous.

To see that $\mathcal{L}_\varphi$ is a left ideal, let $x \in \mathcal{L}_\varphi$ and $y \in \mathcal{A}$. We have

$$\varphi\big((yx)^*(yx)\big) = \varphi\big(x^*(y^*yx)\big) = 0,$$

so that yx is in $\mathcal{L}_\varphi$ and $\mathcal{L}_\varphi$ is a left ideal. ∎

Let φ be a state on a C^*-algebra $\mathcal{A}$. Let

$$H^0_\varphi = \mathcal{A}/\mathcal{L}_\varphi,$$

where the quotient is taken with respect to the linear structure: $x \sim y$ if and only if $x - y \in \mathcal{L}_\varphi$. We denote the cosets in the quotient by $[x]$. It is clear that H^0_φ is a pre-Hilbert space with the inner product

$$\big\langle [x], [y] \big\rangle = \varphi(y^* x), \qquad x,\, y \in \mathcal{A}.$$

We let H_φ denote the completion of H^0_φ with respect to this inner product.

PROPOSITION 14.2

Let φ be a state on a C^*-algebra $\mathcal{A}$. For any $x \in \mathcal{A}$ define an operator $T_x : H^0_\varphi \to H^0_\varphi$ by

$$T_x\big([y]\big) = [xy], \qquad y \in \mathcal{A}.$$

Then T_x is well-defined and extends to a bounded linear operator on H_φ with $\|T_x\| \leq \|x\|$.

PROOF If $y_1 - y_2 \in \mathcal{L}_\varphi$, then $xy_1 - xy_2 \in \mathcal{L}_\varphi$ since $\mathcal{L}_\varphi$ is a left ideal. Thus T_x is well-defined. It is clear that T_x is linear on H^0_φ. It remains to show that

$$\big\|T_x([y])\big\| \leq \|x\| \, \big\|[y]\big\|, \qquad x,\, y \in \mathcal{A}.$$

Fix x and y in $\mathcal{A}$. We have

$$\begin{aligned} \|x\|^2 \big\|[y]\big\|^2 - \big\|T_x([y])\big\|^2 &= \|x\|^2 \varphi(y^* y) - \varphi\big((xy)^*(xy)\big) \\ &= \varphi\big(y^*(\|x\|^2 \mathbf{1} - x^* x) y\big). \end{aligned}$$

Since $\|x\|^2\mathbf{1} - x^*x$ is positive, we can write

$$\|x\|^2 \mathbf{1} - x^* x = \widetilde{x}^* \widetilde{x}$$

for some $\widetilde{x}$ in $\mathcal{A}$ and hence

$$y^*\big(\|x\|^2\mathbf{1} - x^*x\big)y = (\widetilde{x}y)^*(\widetilde{x}y) \geq 0.$$

Since φ is postive, we have

$$\|x\|^2 \big\|[y]\big\|^2 - \big\|T_x([y])\big\|^2 = \varphi\big(y^*(\|x\|^2\mathbf{1} - x^*x)y\big) \geq 0,$$

completing the proof of the proposition. ∎

THEOREM 14.3

Suppose $\mathcal{A}$ is a C^*-algebra and φ is a state on $\mathcal{A}$. The mapping $\Phi : \mathcal{A} \to B(H_\varphi)$ defined by $\Phi(x) = T_x$, $x \in \mathcal{A}$, is a C^*-homomorphism.

PROOF For $x_1, x_2, y \in \mathcal{A}$ and $a, b \in \mathbf{C}$ we have

$$\begin{aligned} T_{ax_1+bx_2}\big([y]\big) &= [ax_1y + bx_2y] \\ &= a[x_1y] + b[x_2y] \\ &= aT_{x_1}\big([y]\big) + bT_{x_2}\big([y]\big). \end{aligned}$$

Since H_φ^0 is dense in H_φ, we have

$$\Phi(ax_1 + bx_2) = a\Phi(x_1) + b\Phi(x_2),$$

so that Φ is linear.

Similarly,

$$T_{x_1x_2}\big([y]\big) = [x_1x_2y] = T_{x_1}\big([x_2y]\big) = T_{x_1}T_{x_2}\big([y]\big)$$

implies that

$$\Phi(x_1x_2) = \Phi(x_1)\Phi(x_2),$$

so that Φ is multiplicative. It is clear that $\Phi(\mathbf{1}) = I$.

Finally, for y_1, y_2, and x in $\mathcal{A}$ we have

$$\begin{aligned} \big\langle T_{x^*}\big([y_1]\big), [y_2]\big\rangle &= \varphi(y_2^*x^*y_1) \\ &= \varphi\big((xy_2)^*y_1\big) = \big\langle [y_1], [xy_2]\big\rangle \\ &= \big\langle [y_1], T_x\big([y_2]\big)\big\rangle = \big\langle T_x^*\big([y_1]\big), [y_2]\big\rangle. \end{aligned}$$

This shows that $\Phi(x^*) = \Phi(x)^*$. ∎

14.2 Review on Direct Sum of Hilbert Spaces

Suppose $\{H_\alpha\}_{\alpha\in A}$ is a family of Hilbert spaces. The direct sum $H = \bigoplus_{\alpha\in A} H_\alpha$ consists of all $x = \{x_\alpha\}_{\alpha\in A}$ ($x_\alpha \in H_\alpha$ for each $\alpha \in A$) such that

$$\|x\|^2 = \sum_{\alpha\in A} \|x_\alpha\|^2 < +\infty.$$

It is well known that $H = \bigoplus_{\alpha\in A} H_\alpha$ is a Hilbert space with the inner product

$$\langle x, y\rangle = \sum_{\alpha\in A} \langle x_\alpha, y_\alpha\rangle, \qquad x,\, y \in H.$$

Suppose T_α is a bounded linear operator on H_α for each $\alpha \in A$. Define

$$\bigoplus_{\alpha\in A} T_\alpha : \bigoplus_{\alpha\in A} H_\alpha \to \bigoplus_{\alpha\in A} H_\alpha$$

by

$$\bigoplus_{\alpha\in A} T_\alpha(\{x_\alpha\}) = \{T_\alpha x_\alpha\}.$$

Then $\bigoplus_{\alpha\in A} T_\alpha$ is linear on $\bigoplus_{\alpha\in A} H_\alpha$ with

$$\left\|\bigoplus_{\alpha\in A} T_\alpha\right\|_H = \sup\{\|T_\alpha\|_{H_\alpha} : \alpha \in A\}.$$

The following properties about the direct sum of bounded linear operators are easy to verify:

$$\bigoplus_{\alpha\in A}(aT_\alpha + bS_\alpha) = a\bigoplus_{\alpha\in A} T_\alpha + b\bigoplus_{\alpha\in A} S_\alpha$$

$$\left(\bigoplus_{\alpha\in A} T_\alpha\right)^* = \bigoplus_{\alpha\in A} T_\alpha^*$$

$$\left(\bigoplus_{\alpha\in A} T_\alpha\right)\left(\bigoplus_{\alpha\in A} S_\alpha\right) = \bigoplus_{\alpha\in A} T_\alpha S_\alpha.$$

14.3 A Representation of C^*-Algebras

We now show that every C^*-algebra can be represented as a C^*-subalgebra of $B(H)$ for some Hilbert space H.

THEOREM 14.4

For every C^*-algebra $\mathcal{A}$ there exists a Hilbert space H such that $\mathcal{A}$ is C^*-isomorphic to a C^*-subalgebra of $B(H)$.

PROOF Recall that $S(\mathcal{A})$ is the space of all states on $\mathcal{A}$. Let

$$H = \bigoplus_{\varphi\in S(\mathcal{A})} H_\varphi,$$

where H_φ are the Hilbert spaces defined in Section 14.1. Define

$$\Phi : \mathcal{A} \to B(H)$$

by

$$\Phi(x) = \bigoplus_{\varphi\in S(\mathcal{A})} T_x^{(\varphi)},$$

where $T_x^{(\varphi)} : H_\varphi \to H_\varphi$ is defined by

$$T_x^{(\varphi)}([y]) = [xy], \qquad y \in \mathcal{A}.$$

By Theorem 14.3 and the properties about direct sums of operators stated in Section 14.2, the mapping Φ is a C^*-homomorphism. Next we show that Φ is one-to-one.

Suppose $\Phi(x) = 0$. Then $T_x^{(\varphi)} = 0$ for every $\varphi \in S(\mathcal{A})$. It follows that $xy \in \mathcal{L}_\varphi$, or

$$\varphi((xy)^*(xy)) = 0, \qquad y \in \mathcal{A},\ \varphi \in S(\mathcal{A}).$$

By Theorem 13.9 we have $(xy)^*(xy) = 0$ and hence $xy = 0$ for all y in $\mathcal{A}$. In particular, $xx^* = 0$, which implies $x = 0$. Therefore, Φ is one-to-one.

Recall from Theorem 10.7 that the image of Φ is a C^*-subalgebra of $B(H)$. Thus Φ is a C^*-isomorphism from $\mathcal{A}$ onto a C^*-subalgebra of $B(H)$, completing the proof of the theorem. ▮

Note that in proving Theorem 14.4 we do not need the whole state space $S(\mathcal{A})$. It suffices to use the space $P(\mathcal{A})$ of pure states when forming the direct sum H. We can also use any set that is between $P(\mathcal{A})$ and $S(\mathcal{A})$. However, if we form the direct sum H using a set that is strictly contained in $P(\mathcal{A})$, then the C^*-homomorphism obtained in the GNS construction may no longer be an isomorphism.

Exercises

14.1 Convince yourself of the properties listed in Section 14.2.

14.2 Give another proof of Theorem 11.5 using representations of C^*-algebras.

14.3 Suppose $\mathcal{A}$ is a C^*-subalgebra of $B(H)$ and x is a unit vector in H.

- (a) Show that $T \mapsto \langle Tx, x\rangle$ is a state for $\mathcal{A}$. Denote this state by φ_x.
- (b) If x is cyclic (see Lecture 22 for definition) for $\mathcal{A}$, then H_{φ_x} can be identified with H in a natural way.
- (c) If x is separating (see Lecture 22 for definition) for $\mathcal{A}$, then H_{φ_x} can be identified with a subspace of H in a natural way.

14.4 Let $\mathcal{A}$ be the commutative C^*-algebra $\mathbf{C}(\partial\mathbf{D})$. For any $z \in \partial\mathbf{D}$ let φ_z be the point-evaluation at z. Show that H_{φ_z} is one-dimensional. Show that there exists a countable subset S' of $S(\mathcal{A})$ such that

$$\Phi : \mathcal{A} \to B\left(\bigoplus_{\varphi \in S'} H_\varphi\right)$$

as constructed in the proof of Theorem 14.4 is a C^*-isomorphism.

15

Non-Unital C^*-Algebras

15.1 Adjoining a Unit

When a "unit" is desired of a non-unital C^*-algebra, one can usually do the following: adjoining a unit or finding an approximate unit. A Banach algebra without unit can easily be embedded in a unital Banach algebra; see Section 7.1. However, embedding a non-unital C^*-algebra into a unital one is a little more complicated.

Suppose $\mathcal{A}$ is a C^*-algebra without identity (or unit). The definition of such algebras should now be obvious. Let

$$\widetilde{\mathcal{A}} = \{(x, a) : x \in \mathcal{A}, a \in \mathbf{C}\}.$$

It is clear that $\widetilde{\mathcal{A}}$ is a linear space with addition and scalar multiplication defined componentwise. Also $\widetilde{\mathcal{A}}$ is an algebra if multiplication is defined by

$$(x, a)(y, b) = (xy + bx + ay, ab), \qquad x, y \in \mathcal{A},\ a, b \in \mathbf{C}.$$

For x in $\mathcal{A}$ and a in $\mathbf{C}$ let $(x, a)^* = (x^*, \bar{a})$. It is easy to check that this defines an involution on $\widetilde{\mathcal{A}}$. However, the natural norm on $\widetilde{\mathcal{A}}$ defined by

$$\|(x, a)\| = \|x\| + |a|$$

is not compatible with the above involution when one tries to make $\widetilde{\mathcal{A}}$ a C^*-algebra. Nevertheless, the following result shows that it is always possible to extend the norm on $\mathcal{A}$ to one on $\widetilde{\mathcal{A}}$ which makes $\widetilde{\mathcal{A}}$ a C^*-algebra.

THEOREM 15.1
Suppose $\mathcal{A}$ is a C^*-algebra without identity. There exists a unique norm on $\widetilde{\mathcal{A}}$ that makes $\widetilde{\mathcal{A}}$ a C^*-algebra and satisfies $\|(x, 0)\| = \|x\|$ for all x in $\mathcal{A}$.

PROOF The uniqueness was proven in Section 8.3. The existence can be proved constructively. Consider the following semi-norm on $\widetilde{\mathcal{A}}$:

$$\big\|(x, a)\big\| = \sup\big\{\|xy + ay\| : y \in \mathcal{A}, \|y\| \le 1\big\}.$$

This is simply the norm of the operator of left multiplication on $\mathcal{A}$ induced by the element (x, a). It is easy to see that $\big\|(x,0)\big\| = \|x\|$ for all x in $\mathcal{A}$. In fact,

$$\big\|(x,0)\big\| = \sup\big\{\|xy\| : y \in \mathcal{A}, \|y\| \le 1\big\} \le \|x\|$$

since $\|xy\| \le \|x\|\,\|y\|$. On the other hand, if $\|x\| \ne 0$, then for $y = x^*/\|x\|$ we have $\|y\| = 1$ and

$$\|xy\| = \frac{\|xx^*\|}{\|x\|} = \|x\|,$$

so that $\big\|(x,0)\big\| \ge \|x\|$ and hence $\big\|(x,0)\big\| = \|x\|$ for all $x \in \mathcal{A}$.

Suppose $x, y \in \mathcal{A}$ and $a, b \in \mathbf{C}$. We have

$$\begin{aligned}\big\|(x,a)(y,b)\big\| &= \big\|(xy + ay + bx, ab)\big\| \\ &= \sup\big\{\|(xy + ay + bx)z + abz\| : z \in \mathcal{A}, \|z\| \le 1\big\} \\ &= \sup\big\{\|x(yz + bz) + a(yz + bz)\| : z \in \mathcal{A}, \|z\| \le 1\big\}.\end{aligned}$$

Since

$$\begin{aligned}\|x(yz + bz) + a(yz + bz)\| &\le \big\|(x,a)\big\|\,\|yz + bz\| \\ &\le \big\|(x,a)\big\|\,\big\|(y,b)\big\|\,\|z\|,\end{aligned}$$

we see that

$$\big\|(x,a)(y,b)\big\| \le \big\|(x,a)\big\|\,\big\|(y,b)\big\|$$

for all (x,a) and (y,b) in $\widetilde{\mathcal{A}}$.

To show that $\|\ \ \|$ as defined above is actually a norm on $\widetilde{\mathcal{A}}$, suppose $\big\|(x,a)\big\| = 0$. If $a = 0$, then $\|x\| = \big\|(x,0)\big\| = 0$ and hence $x = 0$. If $a \ne 0$, we can consider the element $e = -x/a$ in $\mathcal{A}$. Since $xy + ay = 0$ for all $y \in \mathcal{A}$, it is easy to check that $ey = y$ for every y in $\mathcal{A}$, so that e is a left identity for $\mathcal{A}$. Taking adjoints we see that e^* is a right identity for $\mathcal{A}$. Therefore, $e = e^*$ is an indentity for $\mathcal{A}$, contradicting our assumption that $\mathcal{A}$ is non-unital.

Finally we show that

$$\big\|(x,a)\big\|^2 = \big\|(x,a)^*(x,a)\big\|, \qquad x \in \mathcal{A}, a \in \mathbf{C}.$$

Since

$$\big\|(x,a)^*(x,a)\big\| \le \big\|(x,a)\big\|^2,$$

we need only prove

$$\big\|(x,a)\big\|^2 \le \big\|(x,a)^*(x,a)\big\|$$

for all $x \in \mathcal{A}$ and $a \in \mathbf{C}$. Moreover, it suffices to do this for $\|(x,a)\| = 1$. For each $0 < r < 1$ there exists $y \in \mathcal{A}$ such that $\|y\| \leq 1$ and $\|(x,a)(y,0)\| \geq r$. Since each $(x,a)(y,0)$ belongs to $\mathcal{A}$, we have

$$\begin{aligned} \|(x,a)^*(x,a)\| &\geq \|(y,0)^*(x,a)^*(x,a)(y,0)\| \\ &= \left\|[(x,a)(y,0)]^*[(x,a)(y,0)]\right\| \\ &= \|(x,a)(y,0)\|^2 \geq r^2, \end{aligned}$$

and therefore $\|(x,a)^*(x,a)\| \geq 1$. ∎

15.2 Approximate Identities

Suppose $\mathcal{A}$ is a Banach algebra (unital or non-unital). An approximate identity of $\mathcal{A}$ is a net $\{e_\alpha\}$ of elements in $\mathcal{A}$ with the following properties:

(a) $\|e_\alpha\| \leq 1$ for every α.

(b) $\|e_\alpha x - x\| \to 0$ and $\|x e_\alpha - x\| \to 0$ for every $x \in \mathcal{A}$.

Let $\mathcal{A}$ be a C^*-algebra (unital or non-unital). We say that an approximate identity $\{e_\alpha\}$ of $\mathcal{A}$ is increasing if $e_\alpha \geq 0$ for every α and $e_\alpha \leq e_\beta$ whenever $\alpha \leq \beta$. We now establish the existence of approximate identities for non-unital C^*-algebras.

THEOREM 15.2

Suppose $\mathcal{A}$ is a unital C^*-algebra and $\mathcal{I}$ is a left ideal of $\mathcal{A}$. Then there exists an increasing net $\{e_\alpha\}$ of positive elements in the closed unit ball of $\mathcal{I}$ such that $\|x e_\alpha - x\| \to 0$ for every $x \in \mathcal{I}$.

PROOF Let Λ be the set of finite subsets of $\mathcal{I}$ ordered by inclusion. For $\alpha = \{x_1, \cdots, x_n\} \in \Lambda$ let

$$v_\alpha = x_1^* x_1 + \cdots + x_n^* x_n.$$

Since $\mathcal{I}$ is a left ideal of $\mathcal{A}$, each v_α is a positive element in $\mathcal{I}$. By the continuous functional calculus and the assumption that $\mathcal{I}$ is a left ideal, we can define for every $\alpha \in \Lambda$ a positive element in $\mathcal{I}$ as follows:

$$e_\alpha = \left(\frac{1}{n} + v_\alpha\right)^{-1} v_\alpha.$$

Since the function

$$f(t) = \left(\frac{1}{n} + t\right)^{-1} t, \qquad t \in \mathbf{R}^+$$

only takes values between 0 and 1, the continuous functional calculus again shows that $0 \leq e_\alpha \leq \mathbf{1}$ for every $\alpha \in \Lambda$. Note that

$$\sum_{i=1}^{n} \left[(e_\alpha - \mathbf{1})x_i\right]^* \left[(e_\alpha - \mathbf{1})\right] = \frac{1}{n^2}\left(\frac{1}{n} + v_\alpha\right)^{-2} v_\alpha$$

and that

$$\left(\frac{1}{n} + t\right)^{-2} t \leq \frac{n}{4}, \qquad t \in \mathbf{R}^+.$$

We have

$$\sum_{i=1}^{n} \left[(e_\alpha - \mathbf{1})x_i\right]^* \left[(e_\alpha - \mathbf{1})x_i\right] \leq \frac{\mathbf{1}}{4n}.$$

For $i = 1, \cdots, n$ we deduce that

$$\left[(e_\alpha - \mathbf{1})x_i\right]^* \left[(e_\alpha - \mathbf{1})\right] \leq \frac{\mathbf{1}}{4n},$$

from which it follows that

$$\left\|(e_\alpha - \mathbf{1})x_i\right\|^2 \leq \frac{1}{4n}.$$

This implies that $\|e_\alpha x - x\| \to 0$ for every $x \in \mathcal{I}$.

To see that $\{e_\alpha\}$ is increasing, let $\alpha, \beta \in \Lambda$ be such that $\alpha \leq \beta$. We have

$$\alpha = \{x_1, \cdots, x_n\}, \quad \beta = \{x_1, \cdots, x_p\}$$

with $p \geq n$, so that

$$v_\alpha \leq v_\beta, \quad \left(\frac{1}{n} + v_\alpha\right)^{-1} \geq \left(\frac{1}{n} + v_\beta\right)^{-1}.$$

Since

$$\frac{1}{n}\left(\frac{1}{n} + t\right)^{-1} \geq \frac{1}{p}\left(\frac{1}{p} + t\right)^{-1}$$

for any $t \geq 0$, by the continuous functional calculus

$$\frac{1}{n}\left(\frac{1}{n} + v_\beta\right)^{-1} \geq \frac{1}{p}\left(\frac{1}{p} + v_\beta\right)^{-1},$$

and so

$$e_\alpha = \mathbf{1} - \frac{1}{n}\left(\frac{1}{n} + v_\alpha\right)^{-1} \leq \mathbf{1} - \frac{1}{n}\left(\frac{1}{n} + v_\beta\right)^{-1} \leq \mathbf{1} - \frac{1}{p}\left(\frac{1}{p} + v_\beta\right)^{-1} = e_\beta.$$

This completes the proof of the theorem. ∎

COROLLARY 15.3
Suppose $\mathcal{A}$ is a unital C^*-algebra and $\mathcal{I}$ is a right ideal of $\mathcal{A}$. Then there exists an increasing net $\{e_\alpha\}$ of positive elements in the closed unit ball of $\mathcal{I}$ such that $\|xe_\alpha - x\| \to 0$ for every $x \in I$.

PROOF Consider $\mathcal{I}^* = \{x^* : x \in \mathcal{I}\}$ and apply the theorem. ∎

COROLLARY 15.4
Suppose $\mathcal{A}$ is a C^*-algebra without unit. Then there exists an increasing approximate identity $\{e_\alpha\}$ for $\mathcal{A}$.

PROOF Let $\widetilde{\mathcal{A}}$ be the unitization of $\mathcal{A}$ given by Theorem 15.1. Then $\mathcal{A}$ is a self-adjoint two-sided ideal in $\widetilde{\mathcal{A}}$. The desired result then follows from Theorem 15.2. ∎

15.3 Two-Sided Ideals in C^*-Algebras

As an application of approximate identities we prove the following important result about two-sided ideals in C^*-algebras.

THEOREM 15.5
Suppose $\mathcal{I}$ is a closed two-sided ideal in a (unital) C^*-algebra $\mathcal{A}$. Then $\mathcal{I}$ is self-adjoint, that is, $x^* \in \mathcal{I}$ whenever $x \in \mathcal{I}$.

PROOF By the results of the previous subsection, there exists an increasing net $\{e_\alpha\}$ of positive elements in the closed unit ball of $\mathcal{I}$ such that $\|e_\alpha x - x\| \to 0$ for every $x \in \mathcal{I}$. Now if x is in $\mathcal{I}$, then

$$\|x^* - x^* e_\alpha\| = \|x - e_\alpha x\| \to 0,$$

which implies that $x^* \in \mathcal{I}$ because each $x^* e_\alpha$ belongs to the ideal $\mathcal{I}$ and $\mathcal{I}$ is closed. ∎

15.4 Quotients of C^*-Algebras

Suppose $\mathcal{A}$ is a C^*-algebra and $\mathcal{I}$ is a closed two-sided ideal of $\mathcal{A}$. Then $\mathcal{I}$ is self-adjoint and so the quotient algebra $\mathcal{A}/\mathcal{I}$ inherits a natural involution. We proceed to show that this involution is compatible with the quotient norm, so that $\mathcal{A}/\mathcal{I}$ is a C^*-algebra.

LEMMA 15.6

Suppose $\mathcal{A}$ is a unital C^*-algebra and $\mathcal{I}$ is a closed two-sided ideal in $\mathcal{A}$. Let $\{e_\alpha\}$ be an increasing approximate identity for the C^*-algebra $\mathcal{I}$. Then

$$\inf\{\|x-y\| : y \in \mathcal{I}\} = \lim_\alpha \|x - xe_\alpha\|$$

for every x in $\mathcal{A}$.

PROOF For every y in $\mathcal{I}$ we have $e_\alpha y - y \to 0$, so that

$$\begin{aligned}\limsup_\alpha \|x - e_\alpha x\| &= \limsup_\alpha \|x - e_\alpha x - y + e_\alpha y\| \\ &= \limsup_\alpha \big\|(\mathbf{1} - e_\alpha)(x-y)\big\| \\ &\le \|x-y\|.\end{aligned}$$

Taking the infimum over all $y \in \mathcal{I}$ we conclude that

$$\limsup_\alpha \|x - e_\alpha x\| \le \inf\{\|x-y\| : y \in \mathcal{I}\}.$$

On the other hand, it is clear that (since $e_\alpha x \in \mathcal{I}$)

$$\liminf_\alpha \|x - e_\alpha x\| \ge \inf\{\|x-y\| : y \in \mathcal{I}\},$$

thus the desired result follows. ∎

THEOREM 15.7

Suppose $\mathcal{A}$ is a (unital) C^*-algebra and $\mathcal{I}$ is a closed two-sided ideal in $\mathcal{A}$. Then $\mathcal{A}/\mathcal{I}$, endowed with the natural involutive algebraic structure and the quotient norm, is a C^*-algebra .

PROOF Let $x \to [x]$ be the quotient mapping from $\mathcal{A}$ onto $\mathcal{A}/\mathcal{I}$. The natural involution on $\mathcal{A}/\mathcal{I}$ is defined by $[x]^* = [x^*]$; this is well-defined since $\mathcal{I}$ is self-adjoint by Theorem 15.5. It is easy to see that $\mathcal{A}/\mathcal{I}$ becomes an involutive Banach algebra with the quotient norm

$$\big\|[x]\big\| = \inf\{\|x-y\| : y \in \mathcal{I}\}.$$

By Lemma 15.6

$$\big\|[x]\big\| = \lim_\alpha \|x - xe_\alpha\|, \qquad x \in \mathcal{A},$$

where $\{e_\alpha\}$ is an increasing approximate identity for $\mathcal{I}$. We need to show that

$$\big\|[x]\big\|^2 \le \big\|[x][x]^*\big\|$$

for all $x \in \mathcal{A}$.

Fix $x \in \mathcal{A}$ and $y \in \mathcal{I}$. Since $\|e_\alpha\| \leq 1$ and $y - e_\alpha y \to 0$, we have

$$ye_\alpha - e_\alpha y e_\alpha = (y - e_\alpha y)e_\alpha \to 0,$$

so that

$$\begin{aligned}\big\|[x]\big\|^2 &= \lim_\alpha \|x - e_\alpha x\|^2 = \lim_\alpha \big\|(x - e_\alpha x)(x - e_\alpha x)^*\big\| \\ &= \lim_\alpha \|xx^* - xx^* e_\alpha - e_\alpha xx^* + e_\alpha xx^* e_\alpha\| \\ &= \lim_\alpha \|xx^* + y - e_\alpha y - e_\alpha xx^* - xx^* e_\alpha - ye_\alpha + e_\alpha y e_\alpha + e_\alpha xx^* e_\alpha\| \\ &= \lim_\alpha \big\|(\mathbf{1} - e_\alpha)(xx^* + y)(\mathbf{1} - e_\alpha)\big\| \\ &\leq \|xx^* + y\|.\end{aligned}$$

Taking the infimum over all y in $\mathcal{I}$, we obtain

$$\big\|[x]\big\|^2 \leq \big\|[x][x]^*\big\|$$

for all $x \in \mathcal{A}$. ∎

15.5 Non-Unital Commutative C^*-Algebras

Suppose Ω is a locally compact (but not compact) Hausdorff space. The Banach algebra $\mathbf{C}_0(\Omega)$ is a non-unital commutative C^*-algebra if involution is defined by complex conjugation. We show that all non-unital commutative C^*-algebras are of this form. We need the following version of the Stone-Weierstrass approximation theorem.

LEMMA 15.8

Let Ω be a locally compact (but not compact) Hausdorff space. Suppose $\mathcal{A}$ is a closed self-adjoint subalgebra of $\mathbf{C}_0(\Omega)$. If $\mathcal{A}$ separates the points of Ω and for every $x \in \Omega$ there exists $f \in \mathcal{A}$ with $f(x) \neq 0$, then $\mathcal{A} = \mathbf{C}_0(\Omega)$.

PROOF See [8]. ∎

THEOREM 15.9

Suppose $\mathcal{A}$ is a non-unital commutative C^*-algebra and $\mathcal{M}$ is the maximal ideal space of $\mathcal{A}$. Then the Gelfand transform $\Gamma : \mathcal{A} \to \mathbf{C}_0(\mathcal{M})$ is a surjective C^*-isomorphism.

PROOF Let $\mathcal{B}$ be the image of the Gelfand transform in $\mathbf{C}_0(\mathcal{M})$. By the mere definition of the maximal ideal space, $\mathcal{B}$ separates the points of $\mathcal{M}$ and for every $\varphi \in \mathcal{M}$ there exists $x \in \mathcal{A}$ such that $\Gamma(x)(\varphi) = \varphi(x) \neq 0$. By Lemma 15.8, we need only show that Γ is isometric (which implies that $\mathcal{B}$ is closed in $\mathbf{C}_0(\mathcal{M})$) and preserves the involutions (so that $\mathcal{B}$ is self-adjoint in $\mathbf{C}_0(\mathcal{M})$).

We first show that Γ is isometric. Let $\widetilde{\mathcal{A}}$ be the C^*-unitization of $\mathcal{A}$ given in Theorem 15.1. Then $\widetilde{\mathcal{A}}$ is a unital commutative C^*-algebra, so that the Gelfand transform of $\widetilde{\mathcal{A}}$ is a surjective (and isometric) C^*-isomorphism. This implies that for every $x \in \mathcal{A}$ there exists a multiplicative linear functional $\widetilde{\varphi}$ on $\widetilde{\mathcal{A}}$ such that

$$\|x\| = \big\|(x,0)\big\| = \big|\widetilde{\varphi}(x,0)\big|.$$

Let φ be the restriction of $\widetilde{\varphi}$ on $\mathcal{A}$. Then φ belongs to $\mathcal{M}$ and $\big|\varphi(x)\big| = \|x\|$, so that $\big\|\Gamma(x)\big\| \geq \|x\|$. But $\big\|\Gamma(x)\big\| \leq \|x\|$, we must have $\big\|\Gamma(x)\big\| = \|x\|$ for all $x \in \mathcal{A}$.

Next we show that Γ preserves the involutions. Writing $x = x_1 + ix_2$, we only need show that $\Gamma(x)$ is real whenever x is self-adjoint. Suppose x is self-adjoint in $\mathcal{A}$ and $\varphi \in \mathcal{M}$. Let

$$\widetilde{\varphi}(x,a) = \varphi(x) + a, \qquad x \in \mathcal{A},\ a \in \mathbf{C}.$$

Then $\widetilde{\varphi}$ is in the maximal ideal space of $\widetilde{\mathcal{A}}$, so that by the results in the unital case, $\widetilde{\varphi}(x,0) = \varphi(x)$ is real. That is, $\Gamma(x)$ is real whenever x is self-adjoint, completing the proof of the theorem. ∎

Part III

VON NEUMANN ALGEBRAS

16

Strong- and Weak-Operator Topologies

16.1 SO and WO Topologies

We now start the study of von Neumann algebras, a special class of C^*-subalgebras of $B(H)$. The theory of von Neumann algebras is by now highly developed and the subject matter is huge. We will only be able to present the very basics of the theory. Readers interested in more advanced topics are referred to [6].

Von Neumann algebras can be defined either algebraically or topologically. We begin with the weak- and strong-operator topologies on $B(H)$. Throughout this lecture H will be a fixed Hilbert space.

DEFINITION 16.1 *The strong-operator (or SO) topology on $B(H)$ is the locally convex topology generated by the semi-norms $\| \ \|_x$, $x \in H$, where*

$$\|T\|_x = \|Tx\|, \qquad T \in B(H).$$

The weak-operator (or WO) topolopy on $B(H)$ is the locally convex topology generated by the semi-norms $\| \ \|_{x,y}$, $x, y \in H$, where

$$\|T\|_{x,y} = \big|\langle Tx, y\rangle\big|, \qquad T \in B(H).$$

Note that $T_\alpha \to T$ in the strong-operator topology if and only if

$$\|T_\alpha x - Tx\| \to 0, \qquad x \in H.$$

And $T_\alpha \to T$ in the weak-operator topology if and only if

$$\langle T_\alpha x, y\rangle \to \langle Tx, y\rangle, \qquad x, y \in H.$$

By the Cauchy-Schwarz inequality, strong-operator convergence implies weak-operator convergence. When the dimension of H is infinite, the weak-operator topology is strictly weaker than the strong-operator topology. In fact, if T_n is the operator on $L^2(\partial\mathbf{D}, dt)$ defined by

$$T_n f(t) = e^{int} f(t), \qquad f \in L^2(\partial\mathbf{D}, dt),$$

then $T_n \to 0$ in the weak-operator topology. But $T_n \not\to 0$ in the strong-operator topology, since each T_n is unitary.

Suppose $\{T_n\}$ is a sequence of bounded linear operators on H. By the uniform boundedness principle, the weak-operator (and hence the strong-operator) convergence of $\{T_n\}$ implies that there exists a constant $C > 0$ such that $\|T_n\| \leq C$ for every n. This is no longer true when "sequences" are replaced by "nets"; see Exercise 4.3 of [7].

Suppose $T_\alpha \to T$ in the weak-operator topology. It is easy to see that $T_\alpha^* \to T^*$ in the weak-operator topology. The corresponding result fails for the strong-operator topology when $\dim H = \infty$. In fact, if T_n $(n \geq 1)$ is the operator on l^2 defined by

$$T_n(a_1, a_2, \cdots) = (a_n, a_{n+1}, \cdots),$$

then $T_n \to 0$, but $T_n^* \not\to 0$, in the strong-operator topology.

Operator multiplication $(S, T) \mapsto ST$ is separately, but not jointly, continuous in both the weak-operator and the strong-operator topology when $\dim H = \infty$. The "separate" cases follow easily from the definitions. To see the "joint" case in the weak-operator topology, consider the operators T_n on $L^2(\partial\mathbf{D}, dt)$ defined a little earlier and let $S_n = T_n^*$. The "joint" case in the strong-operator topology is more subtle and the construction of a counterexample necessarily requires the use of nets. See the exercise at the end of this lecture and Exercise 4.3 in [7].

16.2 SO and WO Continuous Functionals

In view of the many differences between the strong- and weak-operator topologies, the following result is somewhat surprising. Its consequences are far-reaching in the study of von Neumann algebras.

THEOREM 16.1

Suppose φ is a linear functional on $B(H)$. The following are equivalent:

(a) *φ is WO continuous.*

(b) *φ is SO continuous.*

(c) *There exist vectors $x_1, \cdots, x_n$ and $y_1, \cdots, y_n$ in H such that*

$$\varphi(T) = \sum_{k=1}^{n} \langle Tx_k, y_k \rangle, \qquad T \in B(H).$$

PROOF That (a) implies (b) is obvious. It is also clear that (c) implies (a). We proceed to prove that (b) implies (c).

Suppose φ is SO continuous. By Theorem 1.6 there exist $C > 0$ and $x_1, \cdots, x_n$ in H such that

$$|\varphi(T)| \le C \max_{1\le k\le n} \|Tx_k\| \le C\sqrt{\sum_{k=1}^{n} \|Tx_k\|^2}$$

for all T in $B(H)$. Let

$$K_0 = \{(Tx_1, \cdots, Tx_n) : T \in B(H)\}$$

and let K be the norm closure of K_0 in $H \oplus \cdots \oplus H$. The linear functional $\phi : K_0 \to \mathbf{C}$ defined by

$$\phi(Tx_1, \cdots, Tx_n) = \varphi(T)$$

extends to a bounded linear functional on the Hilbert space K. By the Riesz representation theorem (Theorem 1.1), there exists $(y_1, \cdots, y_n)$ in K such that

$$\varphi(T) = \sum_{k=1}^{n} \langle Tx_k, y_k\rangle, \qquad T \in B(H),$$

completing the proof of the theorem. ▮

THEOREM 16.2

Suppose S is a convex subset of $B(H)$. Then the WO closure of S coincides with the SO closure of S in $B(H)$.

PROOF Since S is convex, Theorem 1.7 implies that an operator T in $B(H)$ is in the WO (or SO) closure of S if and only if there exists a net $\{T_\alpha\}$ in S such that $\varphi(T_\alpha) \to \varphi(T)$ for every WO (or SO) continuous linear functional on $B(H)$. By Theorem 16.1 the set of WO continuous linear functionals on $B(H)$ coincides with the set of SO continuous linear functionals on $B(H)$. Thus T is in the WO closure of S if and only if T is in the SO closure of S. ▮

THEOREM 16.3

The closed unit ball $\big(B(H)\big)_1$ of $B(H)$ is compact in the WO topology.

PROOF For each x and y in H let

$$\mathbf{D}_{x,y} = \{z \in \mathbf{C} : |z| \le \|x\|\,\|y\|\}.$$

Each $\mathbf{D}_{x,y}$ is clearly compact. By Tychonoff's theorem $\prod_{x,y\in H}\mathbf{D}_{x,y}$ is compact with the product topology. Define

$$\Phi : \big(B(H)\big)_1 \to \prod_{x,y\in H}\mathbf{D}_{x,y}$$

by

$$\Phi(T)(x,y) = \langle Tx, y\rangle, \qquad T \in \big(B(H)\big)_1.$$

Just as in the proof of Alaoglu's theorem, we can show that Φ is a homeomorphism from $\big(B(H)\big)_1$ onto its image in $\prod_{x,y\in H}\mathbf{D}_{x,y}$ and this image is closed in the product topology. We omit the routine details. ∎

16.3 Von Neumann Algebras

DEFINITION 16.2 *A C^*-subalgebra $\mathcal{A}$ of $B(H)$ is called a von Neumann algebra if $\mathcal{A}$ is closed in the strong-operator topology.*

By Theorem 16.2 a C^*-subalgebra $\mathcal{A}$ of $B(H)$ is a von Neumann algebra if and only if $\mathcal{A}$ is closed in the weak-operator topology. We shall see later that von Neumann algebras can also be defined in terms of the commutant, a purely algebraic notion.

Von Neumann algebras are easy to come by. The full algebra $B(H)$ is clearly a von Neumann algebra. If $\mathcal{A}$ is a C^*-subalgebra of $B(H)$, then the weak- or strong-operator closure of $\mathcal{A}$ is a von Neumann algebra. To see that we can get nontrivial von Neumann algebras this way, let T be a normal operator and let $\mathcal{A}$ be the C^*-subalgebra generated by T. It is clear that $\mathcal{A}$ is commutative. Since taking the weak-operator closure preserves commutativity, we see that the weak-operator closure of $\mathcal{A}$ in $B(H)$ is a nontrivial von Neumann algebra.

Exercises

16.1 Suppose $\{T_n\}$ and $\{S_n\}$ are two sequences in $B(H)$ converging in the strong-operator topology to T and S, respectively. Show that $\{T_nS_n\}$ converges to TS in the strong-operator topology.

16.2 If $T_\alpha \to T$ and $S_\alpha \to S$ in the strong-operator topology and if $\|S_\alpha\| \le C$ for some constant $C > 0$ and all α, then $S_\alpha T_\alpha \to ST$ in the strong-operator topology.

16.3 Show that the norm $T \mapsto \|T\|$ is not continuous in the strong-operator topology. In particular, it is not continuous in the weak-operator topology.

16.4 Show that if $\mathcal{A}$ is a self-adjoint set of commuting operators in $B(H)$, then the weak-operator closure of $\mathcal{A}$ consists again of mutually commuting operators.

16.5 If H is an infinite dimensional Hilbert space, then the closed unit ball of $B(H)$ is not compact in the strong-operator topology.

16.6 If $\mathcal{A}$ is a von Neumann algebra in $B(H)$, then the closed unit ball of $\mathcal{A}$ is compact in the weak-operator topology and closed in the strong-operator topology.

16.7 Verify the assertions made about the operators T_n in Section 16.1.

17

Existence of Projections

17.1

Recall that an element x in a C^*-algebra $\mathcal{A}$ is called a projection if $x^* = x = x^2$. It is clear that both $\mathbf{1}$ and 0 are projections. For some C^*-algebras $\mathbf{1}$ and 0 are the only projections. For example, if K is a connected compact Hausdorff space, $\mathbf{C}(K)$ does not have projections other than the constant functions 1 and 0. We show in this lecture that every von Neumann algebra possesses a lot of projections.

17.2 Monotone Sequences of Operators

Throughout this lecture H will denote a fixed Hilbert space and I will denote the identity operator on H. For $T, S \in B(H)$ we say that $T \geq S$ or $S \leq T$ if $T - S \geq 0$. A set $\{T_\alpha\}$ is said to be bounded above if there exists T in $B(H)$ such that $T_\alpha \leq T$ for all α. Similarly, $\{T_\alpha\}$ is said to be bounded below if there exists T in $B(H)$ such that $T \leq T_\alpha$ for all α. A net $\{T_\alpha\}$ in $B(H)$ is increasing if $T_\alpha \leq T_\beta$ whenever $\alpha \leq \beta$. Similarly, a net $\{T_\alpha\}$ in $B(H)$ is decreasing if $T_\alpha \geq T_\beta$ whenever $\alpha \leq \beta$.

THEOREM 17.1

Suppose $\mathcal{A}$ is a von Neumann algebra acting on H. If $\{T_\alpha\}$ is an increasing net of self-adjoint operators in $\mathcal{A}$ which is bounded above, then $\{T_\alpha\}$ is strong-operator convergent to a self-adjoint operator T in $\mathcal{A}$. Furthermore, T is the least upper bound of $\{T_\alpha\}$.

PROOF Without loss of generality we may assume that there exists a constant $c > 0$ such that $-cI \leq T_\alpha \leq cI$ for all α. Since $\sigma(T_\alpha) \subset [-c, c]$ and T_α is self-adjoint, $\|T_\alpha\| = r(T_\alpha) \leq c$ for all α. Let

$$\big(B(H)\big)_c = \{T \in B(H) : \|T\| \leq c\}.$$

Then $\big(B(H)\big)_c$ is a compact subset of $B(H)$ in the weak-operator topology by Theorem 16.3. Recall that every net in a compact space has a convergent subnet. Thus $\{T_\alpha\}$ has a subnet which converges in the weak-operator topology. But $\{T_\alpha\}$ is increasing, the convergence of a subnet implies the convergence of the whole net. Thus there exists an operator T in $B(H)$ such that $T_\alpha \to T$ in the weak-operator topology. The limit of every WO convergent net of self-adjoint operators is still self-adjoint, so T is self-adjoint.

Since $\langle T_\beta x, x\rangle \geq \langle T_\alpha x, x\rangle$ for $\beta \geq \alpha$, taking the limit with respect to β leads to $\langle Tx, x\rangle \geq \langle T_\alpha x, x\rangle$ or $T \geq T_\alpha$ for all α, so that T is an upper bound of $\{T_\alpha\}$. If S is another upper bound of $\{T_\alpha\}$, then $\langle Sx, x\rangle \geq \langle T_\alpha x, x\rangle$ for all α and x. Taking the limit with respect to α we get $\langle Sx, x\rangle \geq \langle Tx, x\rangle$ for all x or $S \geq T$. Thus T is the least upper bound of $\{T_\alpha\}$.

Next we show that $T_\alpha \to T$ in the strong-operator topology as well. For each x in H the WO convergence of $\{T_\alpha\}$ gives

$$\big\|(T - T_\alpha)^{\frac{1}{2}} x\big\|^2 = \big\langle (T - T_\alpha)x, x\big\rangle \to 0.$$

Since

$$\big\|(T - T_\alpha)^{\frac{1}{2}}\big\| = \|T - T_\alpha\|^{\frac{1}{2}} \leq (2c)^{\frac{1}{2}}$$

for every α, we have

$$\big\|(T - T_\alpha)x\big\| \leq \big\|(T - T_\alpha)^{\frac{1}{2}}\big\| \, \big\|(T - T_\alpha)^{\frac{1}{2}} x\big\| \to 0$$

for every x in H. So $T_\alpha \to T$ in the strong-operator topology. The operator T remains in $\mathcal{A}$ since $\mathcal{A}$ is strong-operator closed. ∎

COROLLARY 17.2

Suppose $\mathcal{A}$ is a von Neumann algebra and $\{T_\alpha\}$ is a decreasing net of self-adjoint operators in $\mathcal{A}$ which is bounded below. Then $\{T_\alpha\}$ is strong-operator convergent to a self-adjoint operator T in $\mathcal{A}$. Furthermore, T is the greatest lower bound of $\{T_\alpha\}$.

PROOF This follows from Theorem 17.1 by considering $\{-T_\alpha\}$. ∎

COROLLARY 17.3

Suppose $\{P_\alpha\}$ is a net of projections in a von Neumann algebra $\mathcal{A}$ (acting on the Hilbert space H).

(a) *If $\{P_\alpha\}$ is increasing, then it is strong-operator convergent to $P \in \mathcal{A}$, where P is the projection onto the closure of the union of all $P_\alpha H$.*

(b) *If $\{P_\alpha\}$ is decreasing, then it is strong-operator convergent to $P \in \mathcal{A}$, where P is the projection onto the intersection of all $P_\alpha H$.*

PROOF The SO convergence follows from Theorem 17.1, Corollary 17.2, and the fact each projection is between 0 and I. In the first case, the projection P is clearly the least upper bound of $\{P_\alpha\}$; and in the second case, the projection P is clearly the greatest lower bound of $\{P_\alpha\}$. ∎

COROLLARY 17.4

Let $\{P_\alpha\}_{\alpha\in\Lambda}$ be a family of projections in a von Neumann algebra $\mathcal{A}$ (acting on H) whose corresponding subspaces are pairwise orthogonal. Then $\{P_\alpha\}$ is summable in the strong-operator topology to a projection P in $\mathcal{A}$. Moreover,

$$\|Px\|^2 = \sum_\alpha \|P_\alpha x\|^2$$

for every x in H.

PROOF For any finite subset F of Λ let $P_F = \sum_{\alpha\in F} P_\alpha$. Then it is easy to see that $\{P_F\}$ is an increasing net of projections in $\mathcal{A}$. By Corollary 17.3, $\{P_F\}$ is strong-operator convergent to a projection P in $\mathcal{A}$, that is, $\{P_\alpha\}$ is strong-operator summable to P. Furthermore,

$$\|Px\|^2 = \lim_F \|P_F x\|^2 = \lim_F \sum_{\alpha\in F} \|P_\alpha x\|^2 = \sum_{\alpha\in\Lambda} \|P_\alpha x\|^2$$

for every x in H. ∎

17.3 Range Projections

Next we show that a von Neumann algebra contains the range projections of its elements.

THEOREM 17.5

Suppose $\mathcal{A}$ is a von Neumann algebra acting on H and T is in $\mathcal{A}$ with $0 \le T \le I$. Then $\{T^{\frac{1}{n}}\}$ is an increasing sequence whose strong-operator limit is the projection onto the closure of the range of T.

PROOF If $f(t)$ is a continuous function on $\mathbf{R}$ with $0 \le f(t) \le 1$ for all t, it is clear that $\{f^{\frac{1}{n}}\}$ is an increasing sequence of continuous functions bounded above by the constant function 1. By the continuous functional calculus for T the sequence $\{T^{\frac{1}{n}}\}$ is increasing and bounded above by I. By Theorem 17.1 there exists a self-adjoint P in $\mathcal{A}$ such that $T^{\frac{1}{n}} \to P$ in the strong-operator topology. Next we show that P is the projection onto the closure of the range of T.

Since

$$\|(T^{\frac{1}{p}} - P)x\| \to 0, \; \|(T^{\frac{1}{n}} - P)(Px)\| \to 0,$$

and $\|T^{\frac{1}{n}}\| \leq 1$, the identity

$$(T^{\frac{2}{n}} - P^2)x = T^{\frac{1}{n}}(T^{\frac{1}{n}} - P)x + (T^{\frac{1}{n}} - P)(Px)$$

shows that $T^{\frac{2}{n}} \to P^2$ in the strong-operator topology. Passing to the subsequence of even integers we see that $T^{\frac{1}{n}} \to P^2$ in the strong-operator topology. By the uniqueness of the limit, we have $P^* = P = P^2$, so that P is a projection in $\mathcal{A}$.

By the spectral theorem each $T^{\frac{1}{n}}$ is the norm limit of polynomials in T without constant terms. It follows that $T^{\frac{1}{n}}x = 0$ whenever $Tx = 0$. Passing to the SO limit, we see that $Px = 0$ whenever $Tx = 0$. On the other hand, if $Px = 0$, then

$$0 = \langle Px, x\rangle \geq \langle T^{\frac{1}{n}}x, x\rangle = \|T^{\frac{1}{2n}}x\|^2 \geq 0$$

and so $T^{\frac{1}{2n}}x = 0$. Since

$$\|Tx\| \leq \|T^{\frac{2n-1}{2n}}\| \, \|T^{\frac{1}{2n}}x\|,$$

we have $Tx = 0$. Therefore, $\ker P = \ker T$ and P is the orthogonal projection onto the closure of the range of T. ∎

COROLLARY 17.6

Suppose T is an operator in a von Neumann algebra $\mathcal{A}$ acting on H. Then the orthogonal projection onto the closure of the range of T belongs to $\mathcal{A}$.

PROOF Since $TT^*x = 0$ gives

$$\|T^*x\|^2 = \langle TT^*x, x\rangle = 0,$$

we see that

$$\ker T^* = \ker TT^*.$$

This implies that

$$\overline{\operatorname{Ran} T} = \overline{\operatorname{Ran}(TT^*)} = \overline{\operatorname{Ran}\left(\frac{TT^*}{\|TT^*\|}\right)}.$$

Since

$$0 \leq \frac{TT^*}{\|TT^*\|} \leq I,$$

the desired result now follows from Theorem 17.5. ∎

17.4 Existence of Unit

THEOREM 17.7

If $\mathcal{A}$ is a weak-operator closed self-adjoint subalgebra of $B(H)$ (not necessarily containing the identity operator), then there exists a projection P in $\mathcal{A}$, larger than any other projection in $\mathcal{A}$, such that $PT = TP = T$ for all T in $\mathcal{A}$. Thus P is the multiplicative unit for $\mathcal{A}$.

PROOF First observe that all previous results in this lecture remain true without the assumption that the algebra $\mathcal{A}$ contains the identity operator. In particular, if $\mathcal{A}$ is a weak-operator closed self-adjoint subalgebra of $B(H)$, then $\mathcal{A}$ contains all the range projections of operators in $\mathcal{A}$. Let Λ be the net (ordered by inclusion) consisting of all finite subsets of the set of projections in $\mathcal{A}$. For any $\alpha = \{P_1, \cdots, P_n\} \in \Lambda$ let P_α be the projection onto the closure of the range of $P_1 + \cdots + P_n$. It is clear that $\{P_\alpha\}$ is an increasing net of projections bounded above by the projection onto the subspace spanned by all the corresponding subspaces of P_α. By Theorem 17.1, $\{P_\alpha\}$ strong-operator converges to a self-adjoint operator P in $\mathcal{A}$. It is easy to see that P is also a projection. Since P is an upper bound of $\{P_\alpha\}$, P is larger than any other projection in $\mathcal{A}$. In particular, P is larger than the range projection of every operator T, so that $PT = T$ for every T in $\mathcal{A}$. Since $\mathcal{A}$ is self-adjoint, we also have $PT^* = T^*$ or $TP = T$ for every T in $\mathcal{A}$. Thus P is the multiplicative unit for $\mathcal{A}$. ∎

It follows from the theorem above that the requirement in the definition of von Neumann algebras that $\mathcal{A}$ contains the identity operator is not essential. Every weak-operator closed self-adjoint subalgebra of $B(H)$ has a multiplicative unit.

Exercises

17.1 If T is in a von Neumann algebra $\mathcal{A}$, then the orthogonal projection onto the kernel of T belongs to $\mathcal{A}$.

17.2 Suppose $\{P_n\}$ is a sequence of bounded positive operators on H which converges to 0 in the weak-operator topology. Show that $\{P_n\}$ also converges to 0 in the strong-operator topology.

17.3 If $\mathcal{A}$ is a von Neumann algebra of dimension greater than one, show that $\mathcal{A}$ contains nontrivial projections.

18

The Double Commutant Theorem

18.1 The Commutant

In this lecture we show that von Neumann algebras can also be characterized in purely algebraic terms.

DEFINITION 18.1 *Let H be a Hilbert space and let $\mathcal{F}$ be a subset of $B(H)$. The set*

$$\mathcal{F}' = \{T \in B(H) : TS = ST \text{ for all } S \in \mathcal{F}\}$$

is called the commutant of $\mathcal{F}$. The double commutant of $\mathcal{F}$, denoted $\mathcal{F}''$, is simply the commutant of $\mathcal{F}'$.

It is clear that $\mathcal{F}$ is always contained in $\mathcal{F}''$. For a subset $\mathcal{F}$ of $B(H)$ we let $\mathcal{F}^* = \{T^* : T \in \mathcal{F}\}$. We say that $\mathcal{F}$ is self-adjoint if $\mathcal{F}^* = \mathcal{F}$.

PROPOSITION 18.1
$\mathcal{F}'$ is a von Neumann algebra if $\mathcal{F}$ is self-adjoint.

PROOF $\mathcal{F}'$ is clearly a self-adjoint subalgebra of $B(H)$ containing the identity operator. To see that $\mathcal{F}'$ is closed in the weak-operator topology, let $T_\alpha \to T$ with each $T_\alpha \in \mathcal{F}'$. For each S in $\mathcal{F}$ and x, y in H we have

$$\begin{aligned}\langle (TS - ST)x, y\rangle &= \langle TSx, y\rangle - \langle Tx, S^*y\rangle \\ &= \lim_\alpha \big(\langle T_\alpha Sx, y\rangle - \langle T_\alpha x, S^*y\rangle\big) \\ &= \lim_\alpha \langle (T_\alpha S - ST_\alpha)x, y\rangle = 0.\end{aligned}$$

It follows that $TS = ST$ and hence T is in $\mathcal{F}'$. ∎

18.2 Invariant Subspaces

LEMMA 18.2
Suppose H is a Hilbert space and P is the orthogonal projection from H onto a closed subspace M. Then an operator T in $B(H)$ maps M into M if and only if $PTP = TP$.

PROOF Suppose T maps M into M. Then for every $x \in H$ we have $Px \in M$, so that $TPx \in M$ and so $PTPx = TPx$. Thus $PTP = TP$.

On the other hand, if $PTP = TP$, then for every $x \in M$ we have $Px = x$ and so $Tx = TPx = PTPx \in M$. Therefore, T maps M into M. ∎

If T maps a closed subspace M into itself, we say that M is an invariant subspace for T. We say that M is a reducing subspace for T if $TM \subset M$ and $T^*M \subset M$.

The celebrated "invariant subspace problem" is the following: Does every bounded linear operator on an infinite dimensional Hilbert space have a non-trivial invariant subspace?

COROLLARY 18.3
Suppose H is a Hilbert space and P is the orthogonal projection from H onto a closed subspace M. Then $TP = PT$ if and only if M is a reducing subspace for T.

PROOF If $TP = PT$, then clearly $PTP = PT$ and so $TM \subset M$ by Lemma 18.2. Also $TP = PT$ implies that $T^*P = PT^*$. Thus $PT^*P = PT^*$ and $T^*M \subset M$.

If M is a reducing subspace for T, then Lemma 18.2 implies that $PTP = TP$ and $PT^*P = T^*P$. Keeping the first equality and taking the adjoint of the second one, we obtain $PTP = TP$ and $PTP = PT$, so $TP = PT$. ∎

18.3 Von Neumann's Double Commutant Theorem

LEMMA 18.4
Suppose $\mathcal{A}$ is a unital self-adjoint subalgebra of $B(H)$. For each x in H and T in $\mathcal{A}''$ there exists a sequence $\{T_n\}$ in $\mathcal{A}$ such that $\|T_nx - Tx\| \to 0$.

PROOF Let M be the closure in H of the subspace $\{Tx : T \in \mathcal{A}\}$. Since $\mathcal{A}$ is a self-adjoint subalgebra of $B(H)$, M is a reducing subspace for each operator in $\mathcal{A}$. Let P be the orthogonal projection from H onto M. Then $PT = TP$ for every $T \in \mathcal{A}$ and so $P \in \mathcal{A}'$. Since T is in $\mathcal{A}''$, we have $TP = PT$ and so

M is a reducing subspace for T. In particular, $TM \subset M$. Since $\mathcal{A}$ is unital, the vector x belongs to M and so $Tx \in M$. This implies that there exists a sequence $\{T_n\}$ in $\mathcal{A}$ such that $\|T_n x - Tx\| \to 0$. ▮

COROLLARY 18.5

Suppose $\mathcal{A}$ is a unital self-adjoint subalgebra of $B(H)$. Given $x_1, \cdots, x_N$ in H and T in $\mathcal{A}''$ there exists a sequence $\{T_n\}$ in $\mathcal{A}$ such that

$$\lim_{n\to\infty} \|(T_n - T)x_k\| = 0, \qquad 1 \leq k \leq N.$$

PROOF Let $H_N = H \oplus \cdots \oplus H$ (N copies). Define $\Phi : B(H) \to B(H_N)$ by $\Phi(T) = T \oplus \cdots \oplus T$ (N copies). It is easy to check that $\Phi(\mathcal{A})$ is a unital self-adjoint subalgebra of $B(H_N)$ and $\Phi(T) \in \Phi(\mathcal{A})''$. By Lemma 18.4, there exists a sequence $\{\Phi(T_n)\}$, with each T_n belonging to $\mathcal{A}$, such that

$$\left\|(\Phi(T_n) - \Phi(T))x\right\| \to 0, \qquad n \to \infty,$$

where $x = (x_1, \cdots, x_N)$ is in H_N. It follows that

$$\|(T_n - T)x_k\| \to 0, \qquad n \to \infty,$$

for all $1 \leq k \leq N$. ▮

We can now prove the main result of this lecture.

THEOREM 18.6 THE DOUBLE COMMUTANT THEOREM

Suppose $\mathcal{A}$ is a C^*-subalgebra of $B(H)$. Then $\mathcal{A}$ is a von Neumann algebra if and only if $\mathcal{A} = \mathcal{A}''$.

PROOF If $\mathcal{A} = \mathcal{A}''$, then $\mathcal{A}$ is a von Neumann algebra by Proposition 18.1. Suppose $\mathcal{A}$ is a von Neumann algebra, we proceed to show that $\mathcal{A} = \mathcal{A}''$. Since $\mathcal{A} \subset \mathcal{A}''$ and $\mathcal{A}$ is SO closed, it suffices to show that $\mathcal{A}$ is SO dense in $\mathcal{A}''$.

Given T in $\mathcal{A}''$ and a basic SO neighborhood of T of the form

$$\mathcal{U} = \left\{S \in B(H) : \|(S - T)x_k\| < \epsilon, 1 \leq k \leq N\right\}.$$

By Corollary 18.5, there exists a sequence $\{T_n\}$ in $\mathcal{A}$ such that

$$\|(T_n - T)x_k\| \to 0, \qquad n \to \infty$$

for every $1 \leq k \leq N$. It is then clear that $T_n \in \mathcal{U}$ for n large enough. This shows that every SO neighborhood of T contains elements of $\mathcal{A}$. Since T is arbitrary, we see that $\mathcal{A}$ is SO dense in $\mathcal{A}''$. ▮

The following result is also called von Neumann's double commutant theorem; it is clearly equivalent to Theorem 18.6.

THEOREM 18.7
Suppose $\mathcal{A}$ is a unital self-adjoint subalgebra of $B(H)$. Then $\mathcal{A}$ is strong-operator (as well as weak-operator) dense in $\mathcal{A}''$, that is, $\mathcal{A}''$ is the strong-operator (as well as weak-operator) closure of $\mathcal{A}$.

PROOF This is really a consequence of the proof of von Neumann's double commutant theorem. ▮

COROLLARY 18.8
Suppose $\mathcal{F}$ is a self-adjoint subset of $B(H)$. Then $\mathcal{F}''$ is the smallest von Neumann algebra containing $\mathcal{F}$.

PROOF It is clear that $\mathcal{F}''$ is a von Neumann algebra containing $\mathcal{F}$. If $\mathcal{A}$ is another von Neumann algebra such that $\mathcal{F} \subset \mathcal{A}$, then $\mathcal{A}' \subset \mathcal{F}'$ and so $\mathcal{F}'' \subset \mathcal{A}''$. By von Neumann's double commutant theorem, $\mathcal{F}'' \subset \mathcal{A}'' = \mathcal{A}$, so that $\mathcal{F}''$ is the smallest von Neumann algebra containing $\mathcal{F}$. ▮

Let $\mathcal{F}$ be a subset of $B(H)$. Since the intersection of a family of von Neumann algebras is again a von Neumann algebra, there exists a smallest von Neumann algebra containing $\mathcal{F}$. We call this algebra the von Neumann algebra generated by $\mathcal{F}$. It is clearly the intersection of all von Neumann algebras containing $\mathcal{F}$. If $\mathcal{F}$ is self-adjoint, the corollary above shows that $\mathcal{F}''$ is the von Neumann algebra generated by $\mathcal{F}$. In particular, if $\mathcal{F} = \{T\}$ with $T^* = T$, then $\mathcal{F}''$ coincides with the strong-operator (as well as the weak-operator) closure of the set of polynomials in T. Similarly, if T is normal, then $\{T\}''$ is the strong-operator (and the weak-operator) closure of the set of polynomials in T and T^*. If $\mathcal{F}$ is self-adjoint and consists of mutually commuting operators, then $\mathcal{F}''$ is a commutative von Neumann algebra.

18.4 Polar Decomposition Again

THEOREM 18.9
Suppose $\mathcal{A}$ is a von Neumann algebra acting on H and T is in $\mathcal{A}$. If $T = VP$ is the polar decomposition with $\ker V = \ker P$, then V and P both belong to $\mathcal{A}$.

PROOF Since $\mathcal{A}$ is also a C^*-algebra , we know that $P = \sqrt{T^*T}$ belongs to $\mathcal{A}$. To show that V belongs to $\mathcal{A}$, it suffices to show that V belongs to $\mathcal{A}''$ in view of von Neumann's double commutant theorem.

Suppose S is in $\mathcal{A}'$. For each x in H we have

$$SVPx = STx = TSx$$

and

$$VSPx = VPSx = TSx.$$

We see that SV and VS agree on $\overline{\operatorname{Ran} P}$. Since $\overline{\operatorname{Ran} P}^{\perp} = \ker P$, it remains to show that SV and VS agree on $\ker P = \ker V$. Let $x \in \ker P$. Since $SP = PS$, $PSx = SPx = 0$. It follows that

$$S(\ker P) \subset \ker P = \ker V,$$

so that VS is identically zero on $\ker P$. Clearly SV is also identically zero on $\ker P = \ker V$. We conclude that SV and VS agree on $\ker P$. Thus $SV = VS$ and so V is in $\mathcal{A}''$. ∎

If $\mathcal{A}$ is merely a C^*-algebra acting on a Hilbert space H and if $T = VP$ is the polar decomposition with $\ker V = \ker P$ for an element T in $\mathcal{A}$, the partial isometry V does not belong to $\mathcal{A}$ in general. For instance, if $\overline{\mathbf{D}}$ is the closed unit disk with Lebesgue measure and if $\mathcal{A}$ is the C^*-subalgebra of $B(L^2(\overline{\mathbf{D}}))$ consisting of multiplication operators induced by functions in $\mathbf{C}(\overline{\mathbf{D}})$, then the partial isometry in the polar decomposition of the multiplication operator induced by the identity function does not belong to $\mathcal{A}$.

Exercises

18.1 Prove the conclusion made in the last paragraph.

18.2 Let $H = L^2(\overline{\mathbf{D}})$ with Lebesgue measure and let $\mathcal{A}$ be the algebra of multiplication operators on H induced by functions in $\mathbf{C}(\overline{\mathbf{D}})$. Identify the commutant of $\mathcal{A}$.

18.3 If $\mathcal{A}$ is a C^*-subalgebra of $B(H)$ containing all compact operators, then the commutant of $\mathcal{A}$ consists of just constant multiples of the identity operator. In particular, the commutant of $B(H)$ consists just of the constant multiples of the identity.

18.4 Let $\mathcal{A}$ be a C^*-subalgebra of $B(H)$. Let $\mathcal{A}^{(0)} = \mathcal{A}$ and $\mathcal{A}^{(n+1)} = \left(\mathcal{A}^{(n)}\right)'$ for $n \geq 0$. Show that the sequence $\left\{\mathcal{A}^{(n)}\right\}$ contains at most three distinct algebras, namely, $\mathcal{A}$, $\mathcal{A}'$, and $\mathcal{A}''$.

19

The Kaplansky Density Theorem

19.1 More on the Strong-Operator Topology

Recall that a net $\{T_\alpha\}$ of operators in $B(H)$ converges to $T \in B(H)$ in the strong-operator topology if and only if $\|(T_\alpha - T)x\| \to 0$ for every x in H. If $T_\alpha \to T$ in the strong-operator topology, it is clear that $\|T_\alpha x\| \to \|Tx\|$ and $\langle T_\alpha x, y\rangle \to \langle Tx, y\rangle$ for all x and y in H.

PROPOSITION 19.1
The mapping $T \mapsto T^*$ is strong-operator continuous on the set of normal operators in $B(H)$.

PROOF Recall that T is normal in $B(H)$ if and only if $\|Tx\| = \|T^*x\|$ for all x in H. Suppose $T_\alpha \to T$ in the strong-operator topology and suppose T and all T_α are normal. Then for every x in H we have

$$\begin{aligned}\|(T_\alpha^* - T^*)x\|^2 &= \|T_\alpha x\|^2 + \|Tx\|^2 - \langle x, T_\alpha T^* x\rangle - \langle T_\alpha T^* x, x\rangle \\ &\to \|Tx\|^2 + \|Tx\|^2 - \langle x, TT^* x\rangle - \langle TT^* x, x\rangle \\ &= 2\|Tx\|^2 - 2\|T^*x\|^2 = 0.\end{aligned}$$

Thus $T \mapsto T^*$ is strong-operator continuous on the set of normal operators in $B(H)$. ∎

PROPOSITION 19.2
Suppose f is a continuous function on **C**. *Then $T \mapsto f(T)$ is strong-operator continuous on any bounded set of normal operators in $B(H)$.*

PROOF Suppose $T_\alpha \to T$ in the strong-operator topology and suppose T_α and T are all normal with $\|T_\alpha\| \le r$ and $\|T\| \le r$ for all α. Given $\epsilon > 0$ the Stone-Weierstrass approximation theorem implies that there exists a polynomial

p in z and $\bar{z}$ such that

$$\big|f(z) - p(z, \bar{z})\big| < \epsilon$$

for all $|z| \leq r$. By the continuous functional calculus,

$$\big\|f(T_\alpha) - p(T_\alpha, T_\alpha^*)\big\| < \epsilon, \quad \big\|f(T) - p(T, T^*)\big\| < \epsilon,$$

so that

$$\big\|f(T_\alpha)x - p(T_\alpha, T_\alpha^*)x\big\| \leq \epsilon\|x\|, \quad \big\|f(T)x - p(T, T^*)x\big\| \leq \epsilon\|x\|$$

for every x in H. Using Proposition 19.1 and the facts that $\|T_\alpha\| \leq r$ and $\|T\| \leq r$ for all α, we can easily prove that

$$p(T_\alpha, T_\alpha^*) \to p(T, T^*)$$

in the stong-operator topology. Fix any vector x in H. There exists α_0 such that

$$\big\|p(T_\alpha, T_\alpha^*)x - p(T, T^*)x\big\| < \epsilon, \quad \alpha > \alpha_0.$$

It follows from the triangle inequality that

$$\big\|f(T_\alpha)x - f(T)x\big\| \leq \epsilon\|x\| + \epsilon + \epsilon\|x\|, \quad \alpha > \alpha_0,$$

and hence $f(T_\alpha) \to f(T)$ in the strong-operator topology. ∎

19.2 The Cayley Transform

THEOREM 19.3
The mapping $T \mapsto (T - iI)(T + iI)^{-1}$ is strong-operator continuous from the set of self-adjoint operators in $B(H)$ into the set of unitary operators in $B(H)$.

PROOF Recall that if T is self-adjoint then $\sigma(T) \subset \mathbf{R}$. Since the function

$$u(t) = \frac{t - i}{t + i}$$

maps $\mathbf{R}$ into the unit circle, the continuous functional calculus implies that

$$U(T) = (T - iI)(T + iI)^{-1}$$

is unitary whenever T is self-adjoint. Suppose T and all T_α are self-adjoint and $T_\alpha \to T$ in the strong-operator topology. It is clear that

$$\big\|(T_\alpha - T)(T + iI)^{-1}x\big\| \to 0$$

for every x in H. Since $\|(T_\alpha+iI)^{-1}\| \leq 1$ by the continuous functional calculus and

$$U(T_\alpha) - U(T) = 2i(T_\alpha + iI)^{-1}(T_\alpha - T)(T + iI)^{-1},$$

we conclude that $\|U(T_\alpha)x - U(T)x\| \to 0$ for every x in H. ∎

The mapping

$$U(T) = (T - iI)(T + iI)^{-1}$$

is called the Cayley transform. Suppose U is a unitary operator in $B(H)$ whose spectrum does not contain the number 1. By the continuous functional calculus, the operator

$$H(U) = -i(U + I)(U - I)^{-1}$$

is self-adjoint; this is the inverse of the Cayley transform.

COROLLARY 19.4
Suppose f is a real-valued continuous function on $\mathbf{R}$ which vanishes at ∞. Then $T \mapsto f(T)$ is strong-operator continuous on the set of self-adjoint operators in $B(H)$.

PROOF Consider the function

$$g(z) = f\left(-i\frac{z+1}{z-1}\right) = f\big(-i(z+1)(z-1)^{-1}\big), \quad |z| = 1.$$

The assumptions on f imply that g is continuous on the unit circle. Using the Cayley transform and its inverse we can write $f(T) = g(U(T))$ for every self-adjoint operator T in $B(H)$. Since $U(T)$ is strong-operator continuous from the set of self-adjoint operators to the set of unitary operators, and since $g(U)$ is strong-operator continuous from the (bounded) set of unitary operators to the set of self-adjoint operators (by the proof of Proposition 19.2), we conclude that $T \mapsto f(T)$ is strong-operator continuous on the set of self-adjoint operators on H. ∎

19.3 The Kaplansky Density Theorem

THEOREM 19.5 THE KAPLANSKY DENSITY THEOREM
Suppose $\mathcal{A}$ is a C^*-subalgebra of $B(H)$ and $\mathcal{B}$ is the strong-operator closure of $\mathcal{A}$ in $B(H)$.

(a) *The closed unit ball of $\mathcal{B}$ is the strong-operator closure of the closed unit ball of $\mathcal{A}$.*

(b) *If T is self-adjoint in $\mathcal{B}$ and $\|T\| \le 1$, there exists a net $\{T_\alpha\}$ of self-adjoint operators in $\mathcal{A}$ such that $\|T_\alpha\| \le 1$ and $T_\alpha \to T$ in the strong-operator topology.*

(c) *The set of self-adjoint operators in $\mathcal{B}$ is the strong-operator closure of the set of self-adjoint operators in $\mathcal{A}$.*

PROOF We prove (b) first. Suppose T is self-adjoint in $\mathcal{B}$ and $\|T\| \le 1$. There exists a net $\{S_\alpha\}$ in $\mathcal{A}$ that strong-operator converges to T. Let $S'_\alpha = (S_\alpha + S^*_\alpha)/2$. Since strong-operator convergence implies weak-operator convergence, and the mapping $T \mapsto T^*$ is weak-operator continuous on $B(H)$, we see that the net $\{S'_\alpha\}$ of self-adjoint operators in $\mathcal{A}$ weak-operator converges to T. The set of self-adjoint operators in $\mathcal{A}$ is convex, so that its weak- and strong-operator closures coincide (see Theorem 16.2). Thus there exists a net $\{T_\alpha\}$ of self-adjoint operators in $\mathcal{A}$ such that $T_\alpha \to T$ in the strong-operator topology. Consider the real-valued continuous function f on $\mathbf{R}$ defined by

$$f(t) = \begin{cases} t, & |t| \le 1 \\ \dfrac{1}{t}, & |t| \ge 1. \end{cases}$$

By Corollary 19.4, $f(T_\alpha) \to f(T)$ in the strong-operator topology. Since $f(T) = T$ and each $f(T_\alpha)$ is self-adjoint in $\mathcal{A}$ with norm less than or equal to 1, we have proved (b).

It is clear that (c) is a consequence of (b).

We show that (a) is also a consequence of (b). For any subset $\mathcal{S}$ of $B(H)$ let

$$M_2(\mathcal{S}) = \left\{ \begin{pmatrix} A & B \\ C & D \end{pmatrix} : A, B, C, D \in \mathcal{S} \right\},$$

acting on the Hilbert space $H_2 = H \oplus H$. Since $\mathcal{B}$ is the strong-operator closure of $\mathcal{A}$ in $B(H)$, $M_2(\mathcal{B})$ is the strong-operator closure of $M_2(\mathcal{A})$ in $B(H_2)$. Given $\|T\| \le 1$ in $\mathcal{B}$ let

$$\widetilde{T} = \begin{pmatrix} 0 & T \\ T^* & 0 \end{pmatrix}.$$

It is clear that $\widetilde{T}$ is self-adjoint in $M_2(\mathcal{B})$ with $\|\widetilde{T}\| \le 1$. By (b), there exists a net $\{\widetilde{T}_\alpha\}$ of self-adjoint operators in the closed unit ball of $M_2(\mathcal{A})$ such that $\widetilde{T}_\alpha \to \widetilde{T}$ in the strong-operator topology of $B(H_2)$. Write

$$\widetilde{T}_\alpha = \begin{pmatrix} A_\alpha & T_\alpha \\ T^*_\alpha & B_\alpha \end{pmatrix}$$

for each α. Then each T_α is in $\mathcal{A}$ with $\|T_\alpha\| \le 1$ and $T_\alpha \to T$ in the strong-operator topology of $B(H)$. ∎

COROLLARY 19.6

Suppose $\mathcal{A}$ is a C^*-subalgebra of $B(H)$. Then the following are equivalent:

(a) *$\mathcal{A}$ is a von Neumann algebra.*

(b) *The closed unit ball of $\mathcal{A}$ is weak-operator closed.*

(c) *The closed unit ball of $\mathcal{A}$ is strong-operator closed.*

PROOF If $\mathcal{A}$ is a von Neumann algebra, then $\mathcal{A}$ is weak-operator closed and so

$$(\mathcal{A})_1 = \mathcal{A} \cap \left(B(H)\right)_1$$

is weak-operator closed since $\left(B(H)\right)_1$ is weak-operator compact (see Theorem 16.3).

If the closed unit ball of $\mathcal{A}$ is weak-operator closed, then it is also strong-operator closed, since it is convex. See Theorem 16.2.

If $(\mathcal{A})_1$ is strong-operator closed, then by part (b) of the Kaplansky density theorem, $\mathcal{A}$ and its strong-operator closure share the same closed unit ball. This implies $\mathcal{A}$ coincides with its strong-operator closure and hence is a von Neumann algebra. ∎

Since a convex set in $B(H)$ has the same closure relative to the strong- and weak-operator topologies, we see that the Kaplansky density theorem also holds when the strong-operator topology is replaced by the weak-operator topology. We also note that the Kaplansky density theorem still holds if "self-adjoint operators" are replaced by "positive operators" or by "unitary operators".

Exercises

19.1 Prove parts (b) and (c) in the Kaplansky density theorem for positive and unitary operators.

19.2 If $\mathcal{B}$ is the strong-operator closure of $\mathcal{A}$, then $M_2(\mathcal{B})$ is the strong-operator closure of $M_2(\mathcal{A})$.

19.3 A C^*-subalgebra of $B(H)$ is a von Neumann algebra if and only if its closed unit ball is weak-operator compact. The corresponding statement for the strong-operator topology is false.

19.4 If $\Phi : B(H_1) \to B(H_2)$ is an injective C^*-homomorphism and if Φ is weak-operator continuous, show that Φ maps von Neumann algebras to von Neumann algebras.

19.5 Suppose $\mathcal{A}$ is a von Neumann algebra acting on H and P is a projection in $\mathcal{A}$. Show that $P\mathcal{A}P$ is a von Neumann algebra acting on $H_0 = PH$.

20

The Borel Functional Calculus

20.1 Spectral Measures

In this lecture we prove another version of the spectral theorem for normal operators. In doing so we shall be able to extend the continuous functional calculus for normal operators to the so-called Borel functional calculus. We begin with the notion of spectral measure.

DEFINITION 20.1 *Let K be a compact Hausdorff space and let H be a Hilbert space. A spectral measure E on K (relative to H) is a mapping from the σ-algebra of Borel sets of K to the set of projections in $B(H)$ such that*

(a) $E(\emptyset) = 0$, $E(K) = I$.

(b) $E(S_1 \cap S_2) = E(S_1)E(S_2)$ *for all Borel sets S_1 and S_2.*

(c) *For each x, y in H the function*

$$S \mapsto E_{x,y}(S) = \langle E(S)x, y\rangle$$

is a finite regular complex Borel measure on K.

We shall use $B_\infty(K)$ to denote the set of all bounded Borel functions on K. It is easy to check that $B_\infty(K)$ is a C^*-algebra with the sup-norm and pointwise operations.

Suppose E is a spectral measure on K and f is a function in $B_\infty(K)$. It is clear that the function

$$(x, y) \mapsto \int_K f\, dE_{x,y}$$

is a bounded bilinear form on H. By the Riesz representation theorem (Theorem 1.2), there exists a unique bounded linear operator T on H such that

$$\langle Tx, y\rangle = \int_K f\, dE_{x,y}, \qquad x,\, y \in H.$$

We use the following notation to denote the above fact:

$$T = \int_K f\, dE.$$

We can think of the above integral as being convergent in the weak-operator topology of $B(H)$.

PROPOSITION 20.1
Suppose E is a spectral measure on a compact Hausdorff space K (relative to H). The mapping Φ defined by

$$\Phi(f) = \int_K f\, dE$$

is a C^*-homomorphism from $B_\infty(K)$ into $B(H)$.

PROOF It is clear that Φ is linear. The first property in the definiton of spectral measure implies that $\Phi(1) = I$. Since a spectral measure is projection-valued (and projections are self-adjoint), we also have $\Phi(\bar{f}) = \Phi(f)^*$ for all f in $B_\infty(K)$. It remains to show that Φ is multiplicative.

By the second property in the definition of spectral measure, we have

$$\Phi(fg) = \Phi(f)\Phi(g)$$

whenever f and g are characteristic functions of Borel sets in K. By the linearity of Φ we also

$$\Phi(fg) = \Phi(f)\Phi(g)$$

for step functions (finite linear combinations of characteristic functions of Borel sets) f and g. Since the step functions are dense in $B_\infty(K)$, we conclude that

$$\Phi(fg) = \Phi(f)\Phi(g)$$

for all f and g in $B_\infty(K)$. ∎

20.2 The Spectral Theorem—Version II

Recall that if T is a normal operator on a Hilbert space H, then the maximal ideal space of the (commutative) C^*-subalgebra generated by T can be identified with the spectrum of T. The inverse of the Gelfand transform in this case gives the continuous functional calculus. This result was called the spectral theorem in Lecture 10. Here we present another version of the spectral theorem, which will result in a generalization of the continuous functional calculus, namely, the Borel functional calculus.

THEOREM 20.2 THE SPECTRAL THEOREM
Suppose T is a normal operator in $B(H)$. There exists a unique spectral measure E on $\sigma(T)$ such that

$$T = \int_{\sigma(T)} z\, dE,$$

where z is the identity function on $\sigma(T)$.

PROOF For any fixed x and y in H the function $f \mapsto \langle f(T)x, y\rangle$ is a bounded linear functional on $\mathbf{C}(\sigma(T))$ with norm less than or equal to $\|x\|\,\|y\|$. By the Riesz representation theorem (Theorem 1.3), there exists a unique finite regular complex Borel measure $\mu_{x,y}$ on $\sigma(T)$ such that

$$\|\mu_{x,y}\| \leq \|x\|\,\|y\|$$

and

$$\langle f(T)x, y\rangle = \int_{\sigma(T)} f(z)\, d\mu_{x,y}(z), \quad f \in \mathbf{C}(\sigma(T)).$$

It is clear that $\mu_{x,y}$ depends linearly on x and conjugate linearly on y. Furthermore, if f is a bounded Borel function on $\sigma(T)$,

$$(x, y) \mapsto \int_{\sigma(T)} f(z)\, d\mu_{x,y}(z)$$

is a bounded bilinear form on H with norm less than or equal to $\|f\|_\infty$. Applying the Riesz representation theorem (Theorem 1.2) again, we obtain a bounded linear operator, which we denote by $f(T)$, such that

$$\langle f(T)x, y\rangle = \int_{\sigma(T)} f(z)\, d\mu_{x,y}(z), \qquad x,\, y \in H.$$

Note that $f \mapsto f(T)$ is linear and extends the continuous functional calculus for the normal operator T. We show that the mapping $f \mapsto f(T)$ defined above is a C^*-homomorphism from $B_\infty(\sigma(T))$ into $B(H)$.

The definition of $\mu_{x,y}$ shows that $\mu_{x,x}$ is real-valued for every x in H. This implies that $f(T)$ is self-adjoint if f in $B_\infty(\sigma(T))$ is real-valued. Thus $f \mapsto f(T)$ preserves the involutions.

By the continuous functional calculus,

$$\big\langle (fg)(T)x, x\big\rangle = \big\langle f(T)g(T)x, x\big\rangle$$

or

$$\int_{\sigma(T)} fg\, d\mu_{x,x} = \int_{\sigma(T)} g\, d\mu_{x,\bar{f}(T)x}$$

for all f, g in $\mathbf{C}(\sigma(T))$ and x in H. This clearly implies that

$$f\, d\mu_{x,x} = d\mu_{x,\bar{f}(T)x}$$

for all $f \in \mathbf{C}(\sigma(T))$ and hence

$$\int_{\sigma(T)} fg\, d\mu_{x,x} = \int_{\sigma(T)} g\, d\mu_{x,\bar{f}(T)x} = \int_{\sigma(T)} f\, d\mu_{g(T)x,x}$$

for all f in $\mathbf{C}(\sigma(T))$ and g in $B_\infty(\sigma(T))$. It follows that

$$\int_{\sigma(T)} fg\, d\mu_{x,x} = \int_{\sigma(T)} f\, d\mu_{g(T)x,x}$$

or

$$\langle (fg)x, x\rangle = \langle f(T)g(T)x, x\rangle$$

for all f, g in $B_\infty(\sigma(T))$ and x in H. Using polarization, we see that

$$(fg)(T) = f(T)g(T)$$

for all f and g in $B_\infty(\sigma(T))$ and hence $f \mapsto f(T)$ is a C^*-homomorphism from $B_\infty(\sigma(T))$ into $B(H)$.

For each Borel set S in $\sigma(T)$ let $E(S) = \chi_S(T)$, where χ_S is the characteristic function of S. $E(S)$ is clearly a projection. It is obvious that $E(\emptyset) = 0$ and $E(\sigma(T)) = I$. The equality

$$E(S_1 \cap S_2) = E(S_1)E(S_2)$$

follows from $\chi_{S_1 \cap S_2} = \chi_{S_1}\chi_{S_2}$. If S is a Borel set in $\sigma(T)$, then

$$\langle E(S)x, y\rangle = \int_S d\mu_{x,y}$$

and hence $dE_{x,y} = d\mu_{x,y}$ is a finite regular complex Borel measure. Therefore, the function $S \mapsto E(S)$ defined above is a spectral measure.

For every f in $B_\infty(\sigma(T))$ and x, y in H we have

$$\langle f(T)x, y\rangle = \int_{\sigma(T)} f\, d\mu_{x,y} = \int_{\sigma(T)} f\, dE_{x,y},$$

or

$$f(T) = \int_{\sigma(T)} f\, dE.$$

In particular,

$$T = \int_{\sigma(T)} z\, dE.$$

If E' is another spectral measure on $\sigma(T)$ such that

$$T = \int_{\sigma(T)} z\, dE',$$

then

$$\int_{\sigma(T)} f\,dE = \int_{\sigma(T)} f\,dE'$$

for $f = z$ (the identity function on $\sigma(T)$) and $f = 1$ (the constant function on $\sigma(T)$ with value 1). Since these two functions generate the C^*-algebra $\mathbf{C}(\sigma(T))$, it follows from Proposition 20.1 that

$$\int_{\sigma(T)} f\,dE = \int_{\sigma(T)} f\,dE'$$

or

$$\int_{\sigma(T)} f\,dE_{x,y} = \int_{\sigma(T)} f\,dE'_{x,y}$$

for all $f \in \mathbf{C}(\sigma(T))$ and $x, y \in H$. This clearly implies that $E_{x,y} = E'_{x,y}$ for all $x, y \in H$ and hence $E = E'$. ∎

20.3 The Borel Functional Calculus

The spectral measure E constructed in the proof of Theorem 20.2 is called the resolution of identity for the normal operator T.

The C^*-homomorphism $f \mapsto f(T)$ from $B_\infty(\sigma(T))$ into $B(H)$ is called the Borel functional calculus, which extends the continuous functional calculus.

Suppose $\mathcal{A}$ is a von Neumann algebra and T is normal in $\mathcal{A}$. If E is the resolution of identity for T, then for each Borel set S in $\sigma(T)$ the projection $E(S) = \chi_S(T)$ belongs to $\mathcal{A}$. In fact, $f(T)$ belongs to $\mathcal{A}$ for every f in $B_\infty(\sigma(T))$ because

$$\langle f(T)x, y\rangle = \int_{\sigma(T)} f\,dE_{x,y}$$

and so $f(T)$ is in the weak-operator closure of the C^*-subalgebra generated by T.

The following is a nice application of the Borel functional calculus.

THEOREM 20.3

Suppose $\mathcal{A}$ is a von Neumann algebra in $B(H)$. Then the set of finite linear combinations of projections in $\mathcal{A}$ is norm dense in $\mathcal{A}$.

PROOF If T is normal in $\mathcal{A}$, then by the Borel functional calculus T can be approximated in norm by finite linear combinations of projections, because the identity function on $\sigma(T)$ can be approximated in the sup-norm by step functions (finite combinations of characteristic functions). For general $T \in \mathcal{A}$ we can write $T = T_1 + iT_2$ with $T_1, T_2 \in \mathcal{A}$ being self-adjoint. The desired result then follows from the "normal" case. ∎

Exercises

20.1 Let $H = L^2(\overline{\mathbf{D}})$ with Lebesgue measure. Find the resolution of identity for the multiplication operator induced by the identity function.

20.2 Suppose $\mathcal{A}$ is a von Neumann algebra. Then for every positive operator T in the closed unit ball of $\mathcal{A}$ there exists a sequence $\{P_n\}$ of projections in $\mathcal{A}$ such that $T = \sum_{n\geq 1} P_n/2^n$, with the series converging in norm. See [7].

20.3 Let $\mathcal{A}$ be a von Neumann algebra in $B(H)$. Then an operator in $B(H)$ belongs to $\mathcal{A}$ if and only if it commutes with every projection in $\mathcal{A}'$.

20.4 Suppose A is a self-adjoint operator on a Hilbert space H. Then there exists a family $\{P_t\}$ of projections in the von Neumann algebra generated by A, indexed by **R**, such that

(a) $P_t = 0$ if $t < -\|A\|$, and $P_t = I$ if $t > \|A\|$.

(b) $P_t \leq P_s$ if $t \leq s$.

(c)

$$A = \int_{-\|A\|}^{\|A\|} t\, dP_t$$

in the sense of norm convergence of approximating Riemann sums.

(d) A is the norm limit of finite linear combinations with coefficients in $\sigma(A)$ of orthogonal projections $P_s - P_t$.

21

L^∞ as a von Neumann Algebra

21.1 Examples of Abelian von Neumann Algebras

In this and the next lecture we discuss abelian von Neumann algebras. Note that the terms "abelian" and "commutative" are used interchangeably in these lectures. We begin with two canonical examples of abelian von Neumann algebras.

EXAMPLE 21.1

Suppose T is a normal operator on a Hilbert space H and $\mathcal{A}[T]$ is the C^*-subalgebra generated by T and I. We know that $\mathcal{A}[T]$ is abelian. Let $\mathcal{A}_T$ be the strong- or weak-operator closure of $\mathcal{A}[T]$ in $B(H)$. Then $\mathcal{A}_T$ is an abelian von Neumann algebra. Clearly $\mathcal{A}_T$ is the von Neumann algebra generated by $\{T\}$, that is, $\mathcal{A}_T$ is the smallest von Neumann algebra containing T. □

EXAMPLE 21.2

Let K be a compact Hausdorff space and let μ be a finite positive Borel measure on K with $\operatorname{supp}\mu = K$ (This ensures that the inclusion of $\mathbf{C}(K)$ in $L^\infty(K, d\mu)$ is an isometry). For every f in $L^\infty(K, d\mu)$ let M_f be the multiplication operator on $L^2(K, d\mu)$. It is clear that M_f is bounded on $L^2(K, d\mu)$ with $\|M_f\| = \|f\|_\infty$. It is also clear that

$$M_{af+bg} = aM_f + bM_g, \ M_{fg} = M_f M_g, \ M_{\bar{f}} = M_f^*$$

for all $f, g \in L^\infty(K, d\mu)$ and $a, b \in \mathbf{C}$. Thus the mapping $f \mapsto M_f$ is a C^*-isomorphism from $L^\infty(K, d\mu)$ into $B\big(L^2(K, d\mu)\big)$.

Given a compact Hausdorff space K and a finite positive Borel measure μ on K with $\operatorname{supp}\mu = K$ we let

$$\mathcal{C} = \big\{M_f : f \in \mathbf{C}(K)\big\}, \quad \mathcal{L}^\infty = \big\{M_f : f \in L^\infty(K, d\mu)\big\}.$$

Both $\mathcal{C}$ and $\mathcal{L}^\infty$ are abelian C^*-subalgebras of $B\big(L^2(K, d\mu)\big)$. We shall see a little later that $\mathcal{L}^\infty$ is an abelian von Neumann algebra. □

PROPOSITION 21.1

Suppose K is a compact Hausdorff space and μ is a finite positive Borel measure on K with $\operatorname{supp}\mu = K$. *The mapping $f \mapsto M_f$ is a homeomorphism from $L^\infty(K, d\mu)$ onto $\mathcal{L}^\infty$ if $L^\infty(K, d\mu)$, as the dual of $L^1(K, d\mu)$, is equipped with the weak-star topology, and if $\mathcal{L}^\infty$ is equipped with the weak-operator topology.*

PROOF It is easy to see that

$$L^1(K, d\mu) = \big\{g\bar{h} : g, h \in L^2(K, d\mu)\big\}.$$

Thus $f_\alpha \to f$ in the weak-star topology of $L^\infty(K, d\mu)$ if and only if

$$\int_K f_\alpha g\bar{h}\, d\mu \to \int_K fg\bar{h}\, d\mu, \quad g, h \in L^2(K, d\mu),$$

if and only if

$$\langle M_{f_\alpha} g, h\rangle \to \langle M_f g, h\rangle, \quad g, h \in L^2(K, d\mu),$$

if and only if $M_{f_\alpha} \to M_f$ in the weak-operator topology. ∎

The algebra $\mathcal{C}$ is weak-operator dense in $\mathcal{L}^\infty$. In fact, by the theorem above, it suffices to show that $\mathbf{C}(K)$ is weak-star dense in $L^\infty(K, d\mu)$. Since $L^\infty(K, d\mu)$ is the norm closure of simple step functions, it is enough to show that every simple step function can be approximated in the weak-star topology by functions in $\mathbf{C}(K)$. Let

$$f = \sum_{k=1}^n a_k \chi_{E_k}$$

be a simple step function in the closed unit ball of $L^\infty(K, d\mu)$, so that each $|a_k| \le 1$, the E_k's are pairwise disjoint, and the union of the E_k's equals K. Fix $g_1, \cdots, g_m$ in $L^1(K, d\mu)$ and $\epsilon > 0$. There exist compact sets $K_k \subset E_k$ such that

$$\int_{E_i - K_i} |g_j|\, d\mu < \frac{\epsilon}{2n}, \qquad 1 \le i \le n, 1 \le j \le m.$$

By the Tietze extension theorem there exists a function h in $\mathbf{C}(K)$ such that $\|h\|_\infty \le 1$ and $h(x) = a_k$ for x in K_k. It follows that for every $1 \le k \le m$ we have

$$\left|\int_K g_k(f - h)\, d\mu\right| \le \int_K |g_k||f - h|\, d\mu$$

$$= \sum_{i=1}^{n} \int_{E_i - K_i} |g_k||f - h| \, d\mu$$

$$\leq 2 \sum_{i=1}^{n} \int_{E_i - K_i} |g_k| \, d\mu < \epsilon.$$

Therefore, f can be approximated in the weak-star topology by functions in $\mathbf{C}(K)$.

21.2 Maximal Abelian Algebras

PROPOSITION 21.2
A C^*-subalgebra $\mathcal{A}$ of $B(H)$ is a maximal abelian von Neumann algebra if and only if $\mathcal{A} = \mathcal{A}'$.

PROOF If $\mathcal{A} = \mathcal{A}'$, then $\mathcal{A}$ is an abelian von Neumann algebra. Suppose $\widetilde{\mathcal{A}}$ is another abelian von Neumann algebra with $\mathcal{A} \subset \widetilde{\mathcal{A}}$, then $\widetilde{\mathcal{A}} \subset \mathcal{A}' = \mathcal{A}$, so that $\mathcal{A}$ is maximal.

If $\mathcal{A}$ is an abelian von Neumann algebra, $\mathcal{A} \subset \mathcal{A}'$. If $\mathcal{A}$ is also maximal, then $\mathcal{A} = \mathcal{A}'$. In fact, if $T \in \mathcal{A}'$ with $T = T_1 + iT_2$, where T_1 and T_2 are self-adjoint in $\mathcal{A}'$, the von Neumann algebra generated by $\mathcal{A}$ and T_1 is abelian, so that T_1 belongs to $\mathcal{A}$ since $\mathcal{A}$ is maximal. Similarly, T_2 belongs to $\mathcal{A}$ and hence T is in $\mathcal{A}$. ∎

THEOREM 21.3
Let K be a compact Hausdorff space and let μ be a finite positive Borel measure on K with $\operatorname{supp} \mu = K$. Then $\mathcal{L}^\infty$ is a maximal abelian von Neumann algebra in $B(L^2(K, d\mu))$.

PROOF In view of Proposition 21.2 we need only show that $(\mathcal{L}^\infty)' = \mathcal{L}^\infty$. It is clear that $\mathcal{L}^\infty \subset (\mathcal{L}^\infty)'$. Suppose $T \in (\mathcal{L}^\infty)'$. Let $g = T1$. For every f in $L^\infty(K, d\mu)$ we have

$$Tf = TM_f 1 = M_f T1 = M_f g = fg.$$

Since $L^\infty(K, d\mu)$ is dense in $L^2(K, d\mu)$, we see that

$$\sup\{\|fg\|_2 : \|f\|_2 = 1, f \in L^\infty(K, d\mu)\} = \|T\|.$$

It follows that $g \in L^\infty(K, d\mu)$ with $\|g\|_\infty = \|T\|$. Also the equalities

$$Tf = fg = M_g f, \qquad f \in L^\infty(K, d\mu)$$

imply that $T = M_g$ since $L^\infty(K, d\mu)$ is dense in $L^2(K, d\mu)$. ∎

21.3 A Uniqueness Theorem for L^∞

THEOREM 21.4

Suppose μ and ν are finite regular positive Borel measures on a compact metric space K. If there exists a C^*-isomorphism Φ from $L^\infty(K, d\mu)$ onto $L^\infty(K, d\nu)$ which is the identity on $\mathbf{C}(K)$, then μ and ν are mutually absolutely continuous and $L^\infty(K, d\mu) = L^\infty(K, d\nu)$ with Φ being the identity.

PROOF Suppose E is a Borel set in K. Since Φ is a C^*-isomorphism, there exists another Borel set F such that $\Phi(\chi_E) = \chi_F$ (ν-a.e.). When E is closed (hence compact), we can define for each $n \geq 1$ a continuous function on K as follows:

$$\varphi_n(x) = \begin{cases} 1 - nd(x, E), & d(x, E) \leq \frac{1}{n} \\ 0, & d(x, E) \geq \frac{1}{n}, \end{cases}$$

where d is the metric on K and

$$d(x, E) = \inf\{d(x, y) : y \in E\}.$$

It is clear that $\chi_E \leq \varphi_n$ and $\varphi_n(x) \to \chi_E(x)$ pointwise as $n \to +\infty$. Since Φ is order-preserving and preserves functions in $\mathbf{C}(K)$, we have

$$\chi_F = \Phi(\chi_E) \leq \Phi(\varphi_n) = \varphi_n \ (\nu - a.e.)$$

for all $n \geq 1$. Letting $n \to +\infty$ yields $\chi_F \leq \chi_E$ or $F \subset E$ (ν-a.e.).

If E is just a Borel set in K and $\Phi(\chi_E) = \chi_F$ (ν-a.e.), we can use the regularity of μ and ν to find an increasing sequence $\{E_n\}$ of compact sets in K such that

$$E = \bigcup_n E_n \ (\mu - a.e.), \qquad E = \bigcup_n E_n \ (\nu - a.e.).$$

Since Φ is a C^*-isomorphism, it preserves suprema. Thus

$$\chi_F = \Phi(\chi_E) = \Phi(\sup_n \chi_{E_n}) = \sup_n \Phi(\chi_{E_n}) \leq \sup_n \chi_{E_n} = \chi_E \ (\nu - a.e.).$$

Applying this to the compliments of E and F we see that $\Phi(\chi_E) = \chi_E$ $(\nu - a.e.)$ for all Borel sets E in K. This shows that $\mu(E) = 0$ implies $\nu(E) = 0$. By symmetry, we also have $\nu(E) = 0$ implies $\mu(E) = 0$, so that μ and ν are mutually absolutely continuous. Since Φ fixes step functions and those functions are dense in L^∞, we must have

$$L^\infty(K, d\mu) = L^\infty(K, d\nu),$$

and Φ must be the identity mapping. ∎

Exercises

21.1 Show that every C^*-isomorphism preserves suprema.

21.2 Let $\mathcal{L}^\infty$ be the von Neumann algebra of multiplication operators on $L^2(\overline{\mathbf{D}})$ (with Lebesgue measure) induced by bounded functions. Show that $\mathcal{L}^\infty$ has no minimal projections; namely, for every nonzero projection P in $\mathcal{L}^\infty$ there exists another nonzero projection Q in $\mathcal{L}^\infty$ such that $Q \leq P$ but $Q \neq P$.

21.3 Show that the commutant of the algebra $\mathcal{C}$ is equal to $\mathcal{L}^\infty$.

21.4 Let μ and ν be two finite regular positive Borel measures on a compact Hausdorff space. If μ and ν are mutually absolutely continuous, then $L^\infty(K, d\mu) = L^\infty(K, d\nu)$.

22

Abelian von Neumann Algebras

22.1 Cyclic and Separating Vectors

In this lecture we show that every abelian von Neumann algebra acting on a separable Hilbert space is C^*-isomorphic to some $L^\infty(K, d\mu)$. We begin with the important notion of cyclic and separating vectors.

DEFINITION 22.1 *Let $\mathcal{A}$ be a C^*-subalgebra of $B(H)$ and let x be a vector in H. We say that x is a cyclic vector for $\mathcal{A}$ if the set $\{Tx : T \in \mathcal{A}\}$ is dense in H. We say that x is a separating vector for $\mathcal{A}$ if the mapping $T \mapsto Tx$ from $\mathcal{A}$ into H is one-to-one.*

EXAMPLE 22.2

Let $\mathcal{C}$ be the C^*-algebra consisting of multiplication operators M_f on $L^2(\partial\mathbf{D}, dt)$ with $f \in \mathbf{C}(\partial\mathbf{D})$. Then for every integer n the vector $f_n(t) = e^{int}$ is both cyclic and separating for $\mathcal{C}$. ▯

EXAMPLE 22.3

Let $H = H^2(\mathbf{D})$ be the Hardy space of $\mathbf{D}$. If $\mathcal{A}$ is a C^*-subalgebra of $B(H)$ containing M_z (multiplication by the identity function z), then every outer function (see [4] for definiton) in $H^2(\mathbf{D})$ is a cyclic vector for $\mathcal{A}$. ▯

PROPOSITION 22.1

Let $\mathcal{A}$ be a C^*-subalgebra of $B(H)$ and let x be a vector in H. Then x is cyclic for $\mathcal{A}$ if and only if x is separating for $\mathcal{A}'$.

PROOF First assume that x is cyclic for $\mathcal{A}$. If $T \in \mathcal{A}'$ and $Tx = 0$, then for every $S \in \mathcal{A}$ we have $TSx = STx = 0$. Since the set $\{Sx : S \in \mathcal{A}\}$ is dense in H, we must have $T = 0$. Hence x is separating for $\mathcal{A}'$.

Next assume that x is separating for $\mathcal{A}'$. Let $[\mathcal{A}x]$ be the closure of $\{Tx : T \in \mathcal{A}\}$ (note that $x \in [\mathcal{A}x]$ since $I \in \mathcal{A}$) and let P be the orthogonal projection from H onto $[\mathcal{A}x]$. It is easy to see that $[\mathcal{A}x]$ is a reducing subspace for every operator in $\mathcal{A}$. Thus $PS = SP$ for every $S \in \mathcal{A}$ and hence P is in $\mathcal{A}'$. Since $I - P$ is in $\mathcal{A}'$, $(I-P)x = x - x = 0$, and x is separating for $\mathcal{A}'$, we must have $I - P = 0$, or $[\mathcal{A}x] = H$, so that x is cyclic for $\mathcal{A}$. ∎

COROLLARY 22.2

If $\mathcal{A}$ is a commutative C^*-algebra acting on H and $x \in H$ is a cyclic vector for $\mathcal{A}$, then x is also a separating vector for $\mathcal{A}$.

PROOF This follows from Proposition 22.1 and the fact that $\mathcal{A} \subset \mathcal{A}'$ when $\mathcal{A}$ is commutative. ∎

THEOREM 22.3

Suppose $\mathcal{A}$ is a commutative C^*-algebra acting on a separable Hilbert space H. Then $\mathcal{A}$ has a separating vector.

PROOF Let $(H)_1$ be the closed unit ball of H and let

$$\mathcal{E} = \{E \subset (H)_1 : [\mathcal{A}x] \perp [\mathcal{A}y] \text{ for all } x \neq y \text{ in } E\}.$$

Note that $\mathcal{E}$ is nonempty because it contains all singletons in $(H)_1$. Ordering $\mathcal{E}$ by inclusion and applying Zorn's lemma, we obtain a maximal set E_0 in $\mathcal{E}$.

If $y \perp [\mathcal{A}x]$ for every x in E_0, then $[\mathcal{A}y] \perp [\mathcal{A}x]$ for all x in E_0 since $\mathcal{A}$ is a C^*-algebra. This implies that $\{y\} \cup E_0$ is in $\mathcal{E}$, contradicting the maximality of E_0. This shows that

$$H = \bigcup_{x \in E_0} [\mathcal{A}x].$$

Since H is separable, E_0 must be countable, say, $E_0 = \{x_n : n \geq 1\}$.

Let

$$x = \sum_{n=1}^{\infty} \frac{x_n}{2^n}.$$

The vector x is in H since each $\|x_n\| \leq 1$. We show that x is a separating vector for $\mathcal{A}$. Suppose T is in $\mathcal{A}$ and

$$Tx = \sum_{n=1}^{\infty} \frac{Tx_n}{2^n} = 0.$$

Since $Tx_n \perp Tx_m$ for $n \neq m$, we must have $Tx_n = 0$ for each $n \geq 1$. Using the commutativity of $\mathcal{A}$ we conclude that $T[\mathcal{A}x_n] = 0$ for all $n \geq 1$. Since H is the direct sum of $[\mathcal{A}x_n]$ for $n \geq 1$, we see that $T = 0$, and hence x is a separating vector for $\mathcal{A}$. ∎

COROLLARY 22.4
If $\mathcal{A}$ is a maximal abelian von Neumann algebra acting on a separable Hilbert space H, then $\mathcal{A}$ has a cyclic vector.

PROOF Recall that $\mathcal{A}$ is a maximal abelian von Neumann algebra if and only if $\mathcal{A} = \mathcal{A}'$. By the theorem above $\mathcal{A}'(= \mathcal{A})$ has a separating vector x. By Proposition 22.1 x is a cyclic vector for $\mathcal{A}$. ∎

22.2 A Representation of Abelian von Neumann Algebras

THEOREM 22.5
Let $\mathcal{A}$ be an abelian von Neumann algebra in $B(H)$ possessing a (unit) cyclic vector x. There exists a compact Hausdorff space K, a positive Borel measure μ on K with $\operatorname{supp}\mu = K$, *and a unitary $U : H \to L^2(K, d\mu)$ such that $U\mathcal{A}U^*$ coincides with the algebra $\mathcal{L}^\infty$ of all multiplication operators M_f on $L^2(K, d\mu)$ with $f \in L^\infty(K, d\mu)$.*

PROOF Let K be the maximal ideal space of $\mathcal{A}$ and let $\Gamma : \mathcal{A} \to \mathbf{C}(K)$ be the (bijective) Gelfand transform. It is easy to see that $f \mapsto \langle \Gamma^{-1}(f)x, x\rangle$ is a bounded linear functional on $\mathbf{C}(K)$ with norm less than or equal to 1 (recall that x is a unit vector). By the Riesz representation theorem (Theorem 1.3), there exists a unique finite regular Borel measure μ on K such that

$$\langle \Gamma^{-1}(f)x, x\rangle = \int_K f\, d\mu, \qquad f \in \mathbf{C}(K).$$

Since the Gelfand transform preserves positivity, we have

$$\int_K f\, d\mu \geq 0, \qquad f \geq 0,\ f \in \mathbf{C}(K).$$

This clearly implies that μ is positive.

If the support of μ is not the whole space K, there exists a nonempty open set in K whose μ-measure is zero. This implies that there exists a nonzero positive continuous function f in K such that

$$\int_K f\, d\mu = 0, \quad \text{or} \quad \Gamma^{-1}\big(\sqrt{f}\big)x = 0.$$

Since x is separating for $\mathcal{A}$ (see Corollary 22.2), we have $\Gamma^{-1}\big(\sqrt{f}\big) = 0$, so that $\sqrt{f} = 0$ in $\mathbf{C}(K)$, a contradiction. Therefore, $\operatorname{supp}\mu = K$.

For every T in $\mathcal{A}$ we have

$$\int_K |\Gamma(T)|^2\, d\mu = \Big\langle \Gamma^{-1}\big(\Gamma(T^*)\Gamma(T)\big)x, x\Big\rangle = \|Tx\|^2.$$

Thus the mapping $Tx \mapsto \Gamma(T)$ from $\{Tx : T \in \mathcal{A}\}$ to $\mathbf{C}(K) \subset L^2(K, d\mu)$ extends to a unitary U from H onto $L^2(K, d\mu)$. Also for T and S in $\mathcal{A}$ we have

$$UTU^*\big(\Gamma(S)\big) = UTSx = \Gamma(TS) = \Gamma(T)\Gamma(S) = M_{\Gamma(T)}\Gamma(S).$$

Since

$$\big\{\Gamma(S) : S \in \mathcal{A}\big\} = \mathbf{C}(K)$$

is dense in $L^2(K, d\mu)$, we have

$$UTU^* = M_{\Gamma(T)}, \qquad T \in \mathcal{A}.$$

It is easy to see that the mapping $T \mapsto UTU^*$ is a C^*-isomorphism as well as a strong- and weak-operator homeomorphism between $B(H)$ and $B\big(L^2(K, d\mu)\big)$. Thus $U\mathcal{A}U^*$ is a von Neumann algebra in $B\big(L^2(K, d\mu)\big)$. But

$$U\mathcal{A}U^* = \big\{M_f : f \in \mathbf{C}(K)\big\}$$

is weak-operator dense in

$$\mathcal{L}^\infty = \big\{M_f : f \in L^\infty(K, d\mu)\big\}.$$

We conclude that

$$U\mathcal{A}U^* = \big\{M_f : f \in \mathbf{C}(K)\big\} = \big\{M_f : f \in L^\infty(K, d\mu)\big\} = \mathcal{L}^\infty,$$

completing the proof of the theorem. ∎

Note that the space K and the measure $d\mu$ in the proof of the theorem above have a very interesting property. Namely, every function in $L^\infty(K, d\mu)$ can be redefined on a set of measure zero so that the resulting function is in $\mathbf{C}(K)$.

We now prove the main result of this lecture, the representation of abelian von Neumann algebras on separable Hilbert spaces.

THEOREM 22.6

Every abelian von Neumann algebra $\mathcal{A}$ acting on a separable Hilbert space H is C^*-isomorphic to some $L^\infty(K, d\mu)$, where K is a compact Hausdorff space and μ is a finite positive Borel measure on K with $\operatorname{supp} \mu = K$.

PROOF By Theorem 22.3 $\mathcal{A}$ has a (unit) separating vector x in H. Let $H_0 = [\mathcal{A}x]$. H_0 is a reducing subspace for every operator in $\mathcal{A}$. For every T in $\mathcal{A}$ let $\Phi(T)$ be the restriction of T to H_0. It is clear that $\Phi : \mathcal{A} \to B(H_0)$ is linear and multiplicative. Also Φ is one-to-one since x is in H_0 and x is a separating vector for $\mathcal{A}$. If $y_1, y_2 \in H_0$ and $T \in \mathcal{A}$,

$$\big\langle \Phi(T)^* y_1, y_2 \big\rangle = \big\langle y_1, \Phi(T) y_2 \big\rangle = \langle y_1, T y_2 \rangle = \langle T^* y_1, y_2 \rangle = \big\langle \Phi(T^*) y_1, y_2 \big\rangle,$$

so that $\Phi(T^*) = \Phi(T)^*$ and hence Φ is a C^*-isomorphism from $\mathcal{A}$ into $B(H_0)$. In particular, Φ is isometric. Let $\mathcal{B} = \Phi(\mathcal{A})$. Then $\mathcal{B}$ is a C^*-subalgebra of $B(H_0)$ and x is a cyclic vector for $\mathcal{B}$.

Since Φ is isometric, it maps the closed unit ball $(\mathcal{A})_1$ of $\mathcal{A}$ onto the closed unit ball $(\mathcal{B})_1$ of $\mathcal{B}$. Recall that the closed unit ball $\big(B(H)\big)_1$ of $B(H)$ is weak-operator compact and $\mathcal{A}$ is weak-operator closed. This implies that

$$(\mathcal{A})_1 = \mathcal{A} \cap \big(B(H)\big)_1$$

is weak-operator compact. It is easy to see that Φ is weak-operator continuous, so that $(\mathcal{B})_1$ is weak-operator compact. By the Kaplansky density theorem, the weak-operator closure of $\mathcal{B}$ coincides with $\mathcal{B}$, so $\mathcal{B}$ is a von Neumann algebra. The desired result now follows from the previous theorem. ▮

Exercises

22.1 If $\mathcal{A}$ is an abelian von Neumann algebra on a separable Hilbert space, then $\mathcal{A}$ is maximal if and only if $\mathcal{A}$ has a cyclic vector.

22.2 Let H be the Bergman space of **D** consisting of analytic functions in $L^2(\mathbf{D}, dxdy)$. If $\mathcal{A}$ is a commutative C^*-subalgebra of $B(H)$ containing the operator M_z (multiplication by the identity function z), then the vector $f(z) = z - 1$ is both cyclic and separating for $\mathcal{A}$.

22.3 If $\mathcal{A}$ is an abelian C^*-subalgebra of $B(H)$, then there exists a maximal abelian von Neumann algebra in $B(H)$ containing $\mathcal{A}$.

23

The L^∞-Functional Calculus

23.1

Suppose T is a normal operator on a Hilbert space H. Let $\sigma(T)$ be the spectrum of T and let $\mathcal{A}[T]$ be the C^*-algebra generated by T and I. Recall that the Gelfand transform

$$\Gamma : \mathcal{A}[T] \to \mathbf{C}(\sigma(T))$$

is a C^*-isomorphism from $\mathcal{A}[T]$ onto $\mathbf{C}(\sigma(T))$. The inverse of the Gelfand transform is called the continuous functional calculus. In Lecture 20 we were able to extend the continuous functional calculus to the C^*-algebra $B_\infty(\sigma(T))$ of bounded Borel functions on $\sigma(T)$; the result was called the Borel functional calculus. In this lecture we shall extend the Borel functional calculus to the so-called L^∞-functional calculus. Note that

$$B_\infty(\sigma(T)) \subset L^\infty(\sigma(T), d\mu)$$

for any finite Borel measure μ on $\sigma(T)$.

23.2 The L^∞-Functional Calculus

THEOREM 23.1
Suppose T is normal in $B(H)$. If $\mathcal{A}[T]$ has a cyclic vector, there exists a finite positive Borel measure μ on $\sigma(T)$ with $\operatorname{supp}\mu = \sigma(T)$ and a unitary U from H onto $L^2(\sigma(T), d\mu)$ such that $U\mathcal{A}_T U^*$ coincides with the algebra $\mathcal{L}^\infty$ of all multiplication operators M_f on $L^2(\sigma(T), d\mu)$ with $f \in L^\infty(\sigma(T), d\mu)$.

PROOF Recall that the mapping

$$f \mapsto f(T), \qquad f \in \mathbf{C}(\sigma(T))$$

is the continuous functional calculus. Let x be a unit cyclic vector for $\mathcal{A}[T]$. It is clear that $f \mapsto \langle f(T)x, x\rangle$ is a bounded linear functional on $\mathbf{C}(\sigma(T))$ with norm equal to 1. By the Riesz representation theorem (Theorem 1.3) there exists a unique finite regular Borel measure μ on $\sigma(T)$ such that

$$\langle f(T)x, x\rangle = \int_{\sigma(T)} f\, d\mu, \qquad f \in \mathbf{C}(\sigma(T)).$$

Since

$$\int_{\sigma(T)} f\, d\mu \geq 0, \qquad f \geq 0,\ f \in \mathbf{C}(\sigma(T)),$$

the measure μ must be positive. Just as in the proof of Theorem 22.5 we have $\operatorname{supp}\mu = \sigma(T)$.

Let

$$\Gamma : \mathcal{A}[T] \to \mathbf{C}(\sigma(T))$$

be the Gelfand transform. For $S \in \mathcal{A}[T]$ we have

$$\int_{\sigma(T)} |\Gamma(S)|^2\, d\mu = \langle \Gamma(S^*)(T)\Gamma(S)(T)x, x\rangle = \langle S^*Sx, x\rangle = \|Sx\|^2.$$

Since $\{Sx : S \in \mathcal{A}[T]\}$ is dense in H and

$$\{\Gamma(S) : S \in \mathcal{A}[T]\} = \mathbf{C}(\sigma(T))$$

is dense in $L^2(\sigma(T), d\mu)$, the above equality shows that the mapping $Sx \mapsto \Gamma(S)$ extends to a unique unitary U from H onto $L^2(\sigma(T), d\mu)$.

Define

$$\Phi : B(H) \to B(L^2(\sigma(T), d\mu))$$

by

$$\Phi(S) = USU^*, \qquad S \in B(H).$$

It is easy to see that Φ is a bijective C^*-isomorphism as well as an SO (or WO) homeomorphism. If $S_1, S_2 \in \mathcal{A}$, then

$$\begin{aligned}\Phi(S_1)(\Gamma(S_2)) &= US_1U^*(\Gamma(S_2)) = US_1S_2x \\ &= \Gamma(S_1S_2) = M_{\Gamma(S_1)}(\Gamma(S_2)).\end{aligned}$$

Since

$$\{\Gamma(S_2) : S_2 \in \mathcal{A}\} = \mathbf{C}(\sigma(T))$$

is dense in $L^2(\sigma(T), d\mu)$, we have $\Phi(S_1) = M_{\Gamma(S_1)}$. Thus Φ maps $\mathcal{A}[T]$ onto the algebra

$$\mathcal{C} = \{M_f : f \in \mathbf{C}(\sigma(T))\} \subset B(L^2(\sigma(T), d\mu)).$$

Since $\mathcal{A}_T$ is the strong-operator closure of $\mathcal{A}[T]$, $\mathcal{L}^\infty$ is the strong-operator closure of $\mathcal{C}$, and Φ is a strong-operator homeomorphism, we conclude that Φ extends to a unique C^*-isomorphism (as well as an SO or WO homeomorphism) from $\mathcal{A}_T$ onto $\mathcal{L}^\infty$. ∎

The measure μ in the theorem above is unique up to mutually absolute continuity; and the space $L^\infty(\sigma(T), d\mu)$ is unique (independent of μ). In fact, this theorem establishes a C^*-isomorphism from $\mathcal{A}_T$ onto $L^\infty(\sigma(T), d\mu)$ which extends the Gelfand transform

$$\Gamma : \mathcal{A}[T] \to \mathbf{C}(\sigma(T)).$$

Thus if two such measures μ and ν exist, there will be a C^*-isomorphism from $L^\infty(\sigma(T), d\mu)$ onto $L^\infty(\sigma(T), d\nu)$ which is the identity on $\mathbf{C}(\sigma(T))$. By Theorem 21.4, μ and ν are mutually absolutely continuous and

$$L^\infty(\sigma(T), d\mu) = L^\infty(\sigma(T), d\nu).$$

Next we preceed to remove the assumption that $\mathcal{A}$ have a cyclic vector.

THEOREM 23.2

Suppose T is a normal operator on a separable Hilbert space H. There exists a finite positive Borel measure μ on $\sigma(T)$ such that the Gelfand transform

$$\Gamma : \mathcal{A}[T] \to \mathbf{C}(\sigma(T))$$

extends to a unique C^*-isomorphism Γ^* from $\mathcal{A}_T$ onto $L^\infty(\sigma(T), d\mu)$. Furthermore, the measure μ is unique up to mutually absolute continuity and the space $L^\infty(\sigma(T), d\mu)$ is unique (independent of μ).

PROOF Since $\mathcal{A}_T$ is abelian, it has a separating vector. Let x be a unit separating vector for $\mathcal{A}_T$ and let $H_0 = [\mathcal{A}_T x]$ (the closure of $\{Sx : S \in \mathcal{A}_T\}$). Define $\Phi : \mathcal{A}_T \to B(H_0)$ by letting $\Phi(S)$ be the restriction of S to H_0. By the proof of Theorem 22.6, Φ is a C^*-isomorphism from $\mathcal{A}_T$ onto an abelian von Neumann algebra $\mathcal{B}_T$ in $B(H_0)$. It is clear that Φ maps $\mathcal{A}[T]$ onto $\mathcal{B}[\Phi(T)]$, the C^*-subalgebra of $B(H_0)$ generated by $\Phi(T)$ (and the identity on H_0). Since Φ is also strong- (and weak-) operator continuous, we see that $\mathcal{B}_T$ is the von Neumann algebra in $B(H_0)$ generated by $\Phi(T)$ (and the identity on H_0). Recall that a C^*-isomorphism preserves spectra, we have $\sigma(T) = \sigma(\Phi(T))$. Thus if

$$\Gamma' : \mathcal{B}[\Phi(T)] \to \mathbf{C}(\sigma(T))$$

is the Gelfand transform on $\mathcal{B}[\Phi(T)]$, then $\Gamma = \Gamma' \circ \Phi$ must be the Gelfand transform on $\mathcal{A}[T]$.

Since $\Phi(T)$ is normal in $B(H_0)$ and $\mathcal{B}[\Phi(T)]$ has a cyclic vector x, Theorem 23.1 implies that there exists a finite positive Borel measure μ on $\sigma(T)$ such that

$$\Gamma' : \mathcal{B}[\Phi(T)] \to \mathbf{C}(\sigma(T))$$

extends to a C^*-isomorphism from $\mathcal{B}_T$ onto $L^\infty(\sigma(T), d\mu)$. Thus

$$\Gamma = \Gamma' \circ \Phi : \mathcal{A}[T] \to \mathbf{C}(\sigma(T))$$

extends to a C^*-isomorphism from $\mathcal{A}_T$ onto $L^\infty(\sigma(T), d\mu)$. The uniqueness of μ and $L^\infty(\sigma(T), d\mu)$ follows from Theorem 21.4 again. ∎

The mapping $\Gamma^* : \mathcal{A}_T \to L^\infty(\sigma(T), d\mu)$ produced in the proof of the proceeding theorem will be called the extended Gelfand transform. Its inverse is called the L^∞-functional calculus. Since $B_\infty(\sigma(T)) \subset L^\infty(\sigma(T), d\mu)$, the L^∞-functional calculus extends the Borel functional calculus.

23.3 Connectedness of the Unitary Group

As the first application of the L^∞-functional calculus, we extend the definition of spectral integrals to L^∞. Specifically, let T be a normal operator in a von Neumann algebra $\mathcal{A}$. Recall from Lecture 20 that there exists a unique spectral measure E on $\sigma(T)$ such that

$$T = \int_{\sigma(T)} z \, dE.$$

It was also shown in Lecture 20 that the integral

$$f(T) = \int_{\sigma(T)} f \, dE$$

can be defined for every bounded Borel function f on $\sigma(T)$. Now using the L^∞-functional calculus we can define the above integral for an even broader class of functions. In fact, if $d\mu$ is the (more or less unique) positive Borel measure on $\sigma(T)$ given in Theorem 23.2, then for every $f \in L^\infty(\sigma(T), d\mu)$ we can define the integral

$$\int_{\sigma(T)} f \, dE$$

as the operator $f(T)$ from the L^∞-functional calculus.

The following result is another application of the L^∞-functional calculus; it can also be proved using only the Borel functional calculus.

THEOREM 23.3
Suppose $\mathcal{A}$ is a von Neumann algebra. Then the unitary group of $\mathcal{A}$ (that is, the multiplicative group of all unitary operators in $\mathcal{A}$) is path connected in the norm topology.

PROOF Suppose T is unitary in $\mathcal{A}$. In particular, T is normal. Thus there exists a finite positive Borel measure $d\mu$ on $\sigma(T) \subset \partial\mathbf{D}$ such that $f \mapsto f(T)$ is a C^*-isomorphism from $L^\infty(\sigma(T), d\mu)$ onto $\mathcal{A}_T$, the von Neumann subalgebra generated by T.

For $0 \le s \le 1$ let

$$f_s(e^{it}) = e^{ist}, \qquad z = e^{it} \in \partial\mathbf{D},\ 0 \le t < 2\pi.$$

Since f_s is a unimodulus function in $L^\infty(\sigma(T), d\mu)$, the operator $f_s(T)$ is unitary for every $s \in [0,1]$. Moreover, $f \mapsto f_s$ is continuous from $[0,1]$ into $L^\infty(\sigma(T), d\mu)$. Thus by the L^∞-functional calculus the mapping $s \mapsto f_s(T)$ is a norm continuous path in the unitary group of $\mathcal{A}$ connecting T to the identity operator. ∎

COROLLARY 23.4
The invertible group of a von Neumann algebra is path connected in the norm topology.

PROOF Let $\mathcal{A}$ be a von Neumann algebra and let T be an invertible operator in $\mathcal{A}$. By polar decomposition, $T = PU$, where $P \in \mathcal{A}$ is positive and invertible and $U \in \mathcal{A}$ is unitary. By Theorem 23.3, there exists a norm continuous path $t \mapsto U_t$ in the unitary group that connects I to U. Let

$$P_t = (1-t)I + tP$$

and let

$$T_t = P_t U_t, \qquad 0 \le t \le 1.$$

Then $t \mapsto T_t$ is a norm continuous path in the invertible group that connects I to T. ∎

Applying the proceeding corollary to the von Neumann algebra $B(H)$, we obtain

COROLLARY 23.5
Let H be a Hilbert space. Then both the unitary group and the invertible group of $B(H)$ are path connected in the norm topology. In particular, the abstract index of the Banach algebra $B(H)$ is trivial.

Exercises

23.1 Suppose T and S are normal on a separable Hilbert space. If there exists an invertible operator R such that $T = R^{-1}SR$, then there exists a unitary operator U such that $T = U^*SU$.

23.2 Let $H = L^2[-1, 1]$ with Lebesgue measure and let T be the multiplication operator on H induced by the function $f(t) = t^2$. Demonstrate the L^∞-functional calculus for the operator T.

24

Equivalence of Projections

24.1 The Projection Lattice

Suppose $\{P_\alpha\}$ is a family of projections in a von Neumann algebra $\mathcal{A}$ acting on a Hilbert space H. Let M_α be the range of each P_α and let $M = \bigcap_\alpha M_\alpha$. M is clearly a closed subspace of H. We let $\bigwedge_\alpha P_\alpha$ denote the orthogonal projection from H onto M; it is the greatest lower bound of $\{P_\alpha\}$. Let

$$\bigvee_\alpha P_\alpha = I - \bigwedge_\alpha (I - P_\alpha);$$

it is the least upper bound of $\{P_\alpha\}$. It is clear that $\bigvee_\alpha P_\alpha$ is the orthogonal projection from H onto the closed subspace spanned by all M_α. When all M_α are mutually orthogonal, we clearly have

$$\bigvee_\alpha P_\alpha = \sum_\alpha P_\alpha,$$

with the series convergent in the strong-operator topology.

PROPOSITION 24.1
Suppose $\mathcal{A}$ is a von Neumann algebra and $\{P_\alpha\}$ is a family of projections in $\mathcal{A}$. Then $\bigwedge_\alpha P_\alpha$ and $\bigvee_\alpha P_\alpha$ both belong to $\mathcal{A}$.

PROOF Let $P = \bigwedge_\alpha P_\alpha$. Let M_α be the range of each P_α and let $M = \bigcap_\alpha M_\alpha$. For every T in $\mathcal{A}'$ we have $P_\alpha T = TP_\alpha$, so that by Corollary 18.3 $T : M_\alpha \to M_\alpha$ and $T^* : M_\alpha \to M_\alpha$. It follows that $T : M \to M$ and $T^* : M \to M$, or $TP = PT$. By the double commutant theorem, $P \in \mathcal{A}'' = \mathcal{A}$. Since $\mathcal{A}$ contains the identity operator I, we also have $\bigvee_\alpha P_\alpha \in \mathcal{A}$. ∎

COROLLARY 24.2
The set of projections in a von Neumann algebra $\mathcal{A}$ is a complete lattice.

24.2 An Equivalence Relation

The structure of a von Neumann algebra is determined to a large degree by its projection lattice. The study of the projection lattice begins with certain notions of comparing two projections, which we introduce in this lecture and the next.

DEFINITION 24.1 *Two projections P and Q in a von Neumann algebra $\mathcal{A}$ are equivalent, denoted $P \sim Q$, if there exists an operator V in $\mathcal{A}$ such that $V^*V = P$ and $VV^* = Q$.*

If $\mathcal{A}$ acts on the Hilbert space H and $P \sim Q$ via $V^*V = P$, $VV^* = Q$, and $V \in \mathcal{A}$, then by Proposition 12.6 and Corollary 12.7 V is the partial isometry from the range of P onto the range of Q, and V^* is the partial isometry from the range of Q onto the range of P. Thus the range of P is the initial space of V and the range of Q is the final space of V. It is also clear that $Q = VPV^*$ and $P = V^*QV$.

Note that if $\mathcal{A} = B(H)$, then $P \sim Q$ in $\mathcal{A}$ if and only if the corresponding subspaces have the same dimension. If $\mathcal{A}$ is abelian, then clearly $P \sim Q$ in $\mathcal{A}$ if and only if $P = Q$.

PROPOSITION 24.3

The relation $\sim$ is an equivalence relation on the projection lattice of any von Neumann algebra.

PROOF Let $\mathcal{A}$ be a von Neumann algebra and $P \sim Q$ via $V^*V = P$, $VV^* = Q$, and $V \in \mathcal{A}$. It is clear that $Q \sim P$ via $U^*U = Q$, $UU^* = P$, and $U = V^* \in \mathcal{A}$. For any projection P in $\mathcal{A}$ we have $P \sim P$ since $P = P^*P$ and $P = PP^*$. If $P \sim Q$ and $Q \sim R$, there exist U and V in $\mathcal{A}$ such that

$$P = U^*U, \quad Q = UU^*, \qquad Q = V^*V, \quad R = VV^*.$$

Let $W = VU \in \mathcal{A}$. Then

$$W^*W = U^*V^*VU = U^*QU = U^*UU^*U = PP = P$$

and

$$WW^* = VUU^*V^* = VQV^* = VV^*VV^* = RR = R,$$

so that $P \sim R$. Therefore, $\sim$ is an equivalence relation on the projection lattice of $\mathcal{A}$. ∎

24.3 Kaplansky's Formula

LEMMA 24.4
Suppose $\mathcal{A}$ is a von Neumann algebra and T is in $\mathcal{A}$. Let $R(T)$ (respectively $R(T^*)$) be the orthogonal projection onto the closure of the range of T (respectively T^*). Then $R(T) \sim R(T^*)$ in $\mathcal{A}$.

PROOF By Corollary 17.6, both $R(T)$ and $R(T^*)$ belong to $\mathcal{A}$. Also recall from the proof of the polar decomposition $T = V|T|$ that V is the partial isometry from $R(T^*)$ onto $R(T)$. By Theorem 18.4 the partial isometry V belongs to $\mathcal{A}$, so that $R(T) \sim R(T^*)$. ∎

THEOREM 24.5 KAPLANSKY'S FORMULA
If P and Q are projections in a von Neumann algebra $\mathcal{A}$ acting on H, then

$$P \vee Q - Q \sim P - P \wedge Q.$$

PROOF Let M be the range of P and N the range of Q. Since

$$\text{Ran}\,\big((I-Q)P\big)^{\perp} = \ker P(I-Q) = N \oplus (N^{\perp} \cap M^{\perp}),$$

we have

$$I - R\big((I-Q)P\big) = Q + (I-P) \cap (I-Q) = Q + I - P \vee Q,$$

or

$$R\big((I-Q)P\big) = P \vee Q - Q.$$

Similarly,

$$P - P \wedge Q = R\big(P(I-Q)\big) = R\big([(I-Q)P]^*\big).$$

The desired result now follows from Lemma 24.4. ∎

24.4 The Center of a von Neumann Algebra

The center of a von Neumann algebra $\mathcal{A}$ is the set of all operators T in $\mathcal{A}$ which commute with every other operator in $\mathcal{A}$. It is clear that the center of $\mathcal{A}$ is $\mathcal{A} \cap \mathcal{A}'$, which is a von Neumann algebra by itself. A projection in the center $\mathcal{A} \cap \mathcal{A}'$ will be called a central projection of $\mathcal{A}$.

DEFINITION 24.2 *A von Neumann algebra $\mathcal{A}$ will be called a factor if its center is trivial, that is, $\mathcal{A} \cap \mathcal{A}' = \mathbf{C}I$.*

The simplest factor is the full algebra $B(H)$. A factor is necessarily non-abelian (unless the algebra is one-dimensional).

24.5 The Central Support of a Projection

Suppose $\mathcal{A}$ is a von Neumann algebra and P is a projection in $\mathcal{A}$. It is easy to see that there exists a smallest central projection in $\mathcal{A}$ which majarizes P. (We say that Q majarizes P if the range of Q contains the range of P, or $P \leq Q$.) In fact, the set $\{P_\alpha\}$ of all central projections in $\mathcal{A}$ majarizing P is nonempty (it contains I for example), so that $\bigwedge_\alpha P_\alpha$ is the smallest central projection majarizing P. We call this central projection the central support of P and denote it by $Z(P)$.

PROPOSITION 24.6

Let P be a projection in a von Neumann algebra $\mathcal{A}$. Then

$$Z(P) = \bigvee_{T \in \mathcal{A}} R(TP).$$

Recall that $R(S)$ is the orthogonal projection onto the closure of the range of S.

PROOF Let

$$Q = \bigvee_{T \in \mathcal{A}} R(TP).$$

For every operator T in $\mathcal{A}$, we have $Z(P)T = TZ(P)$, so that the range of $Z(P)$ is invariant for T. Since $P \leq Z(P)$, T maps the range of P into the range of $Z(P)$. It follows that $R(TP) \leq Z(P)$ for all $T \in \mathcal{A}$, and hence

$$Q = \bigvee_{T \in \mathcal{A}} R(TP) \leq Z(P).$$

On the other hand, the range of Q is reducing for every operator in $\mathcal{A}$, so that $QT = TQ$ for $T \in \mathcal{A}$. Thus Q is a central projection majarizing P. It follows that $Z(P) \leq Q$ and hence $Z(P) = Q$. ∎

PROPOSITION 24.7

Let P and Q be two projections in a von Neumann algebra $\mathcal{A}$ acting on H. The following are equivalent:

(a) $Z(P)Z(Q) \neq 0$.

(b) $P\mathcal{A}Q \neq \{0\}$.

(c) *There exist nonzero projections $P_1 \leq P$ and $Q_1 \leq Q$ such that $P_1 \sim Q_1$.*

PROOF We first prove that (c) implies (b). Let V be the partial isometry (in $\mathcal{A}$) from the range of Q_1 to the range of P_1, so that $V^*V = Q_1$ and $VV^* = P_1$. Let x be a nonzero vector in the range of Q_1. Then

$$\|PVQx\| = \|PVx\| = \|Vx\| = \|x\| \neq 0.$$

This shows that $PVQ \neq 0$ and hence $P\mathcal{A}Q \neq \{0\}$.

To prove (b) implies (a), observe that for every T in $\mathcal{A}$

$$PTQZ(P)Z(Q) = Z(P)PTQZ(Q) = PTQ.$$

This clearly shows that if $PTQ \neq 0$ for some $T \in \mathcal{A}$ then $Z(P)Z(Q) \neq 0$.

Next we show that (a) implies (c). Suppose $Z(P)Z(Q) \neq 0$ and recall that

$$Z(P) = \bigvee_{T\in\mathcal{A}} R(TP), \qquad Z(Q) = \bigvee_{T\in\mathcal{A}} R(TQ).$$

There exist $T, S \in \mathcal{A}$ and $x, y \in H$ such that

$$\langle TPx, SQy\rangle \neq 0 \quad \text{or} \quad \langle QS^*TPx, y\rangle \neq 0.$$

Let $\widetilde{T} = S^*T$. We have $\widetilde{T} \neq 0$ and $Q\widetilde{T}P \neq 0$. Let

$$Q_1 = R(Q\widetilde{T}P), \qquad P_1 = R(P\widetilde{T}^*Q).$$

It is clear that $Q_1 \leq Q$, $P_1 \leq P$, and $Q_1 \sim P_1$ by Lemma 24.4. ∎

Exercises

24.1 If $\{P_\alpha\}$ and $\{Q_\alpha\}$ are two families of orthogonal projections in a von Neumann algebra such that $P_\alpha \sim Q_\alpha$ for every α, show that $\oplus_\alpha P_\alpha \sim \oplus_\alpha Q_\alpha$.

24.2 If $P \sim Q$ and Z is central, show that $PZ \sim QZ$.

24.3 Let $\mathcal{A} = B(H)$. Then $P \sim Q$ in $\mathcal{A}$ if and only if the dimensions of the range of P and the range of Q are equal.

24.4 If P and Q are projections on H such that $\|P - Q\| < 1$, then the subspaces that correspond to P and Q have the same dimension.

25

A Partial Ordering

25.1 The Definition

DEFINITION 25.1 *Suppose P and Q are projections in a von Neumann algebra $\mathcal{A}$. If there exists a projection Q_1 in $\mathcal{A}$ such that $P \sim Q_1 \leq Q$, we write $P \preceq Q$ or $Q \succeq P$. When $P \preceq Q$ but $P \not\sim Q$, we write $P \prec Q$.*

Note that if $\mathcal{A} = B(H)$ and P, Q are projections in $\mathcal{A}$, then $P \preceq Q$ if and only if the dimension of the range of P is less than or equal to the dimension of the range of Q. If $\mathcal{A}$ is abelian, then $P \preceq Q$ in $\mathcal{A}$ if and only if $P \leq Q$, that is, the range of P is contained in that of Q.

25.2 Checking the Axioms

We proceed to show that the relation $\preceq$ is a partial ordering on the set of equivalence classes (induced by the relation $\sim$ defined in the previous lecture) of the projection lattice of a von Neumann algebra.

LEMMA 25.1
Suppose P and Q are projections in a von Neumann algebra $\mathcal{A}$. If $P \preceq Q$ and $Q \preceq P$, then $P \sim Q$.

PROOF Suppose U and V are partial isometries in $\mathcal{A}$ such that

$$P = U^*U, \qquad UU^* = Q_1 \leq Q$$

and

$$Q = V^*V, \qquad VV^* = P_1 \leq P.$$

Define two sequences $\{P_n\}$ and $\{Q_n\}$ of projections in $\mathcal{A}$ by induction as follows.

$$P_{n+1} = VQ_nV^*, \qquad Q_{n+1} = UP_nU^*, \qquad n \geq 1.$$

It is easy to check that

$$P_2P_1 = P_1P_2 = P_2, \qquad Q_2Q_1 = Q_1Q_2 = Q_2,$$

so that $P_1 \geq P_2$ and $Q_1 \geq Q_2$. Using induction we can verify that both $\{P_n\}$ and $\{Q_n\}$ are decreasing. Let $P_0 = P$, $Q_0 = Q$, and

$$P_\infty = \bigwedge_{n\geq 1} P_n, \qquad Q_\infty = \bigwedge_{n\geq 1} Q_n.$$

Then P_∞ and Q_∞ are in $\mathcal{A}$ and we have the following direct sums

$$P = \sum_{n=0}^{\infty}(P_n - P_{n+1}) + P_\infty, \quad Q = \sum_{n=0}^{\infty}(Q_n - Q_{n+1}) + Q_\infty.$$

Since

$$U(P_n - P_{n+1})U^* = Q_{n+1} - Q_{n+2}, \quad V(Q_n - Q_{n+1})V^* = P_{n+1} - P_{n+2},$$

we see that

$$P_{2n} - P_{2n+1} \sim Q_{2n+1} - Q_{2n+2}, \quad P_{2n+1} - P_{2n+2} \sim Q_{2n} - Q_{2n+1}.$$

Also, by taking the strong-operator limit in

$$P_{n+1} = VQ_nV^*, \qquad Q_{n+1} = UP_nU^*,$$

we see that

$$UP_\infty U^* = Q_\infty, \qquad VQ_\infty V^* = P_\infty,$$

and hence $P_\infty \sim Q_\infty$. Note that forming the direct sum preserves the equivalence (simply take the strong-operator sum of the partial isometries). We then have

$$\begin{aligned} P &= \sum_{n=0}^{\infty}(P_{2n} - P_{2n+1}) + \sum_{n=0}^{\infty}(P_{2n+1} - P_{2n+2}) + P_\infty \\ &\sim \sum_{n=0}^{\infty}(Q_{2n+1} - Q_{2n+2}) + \sum_{n=0}^{\infty}(Q_{2n} - Q_{2n+1}) + Q_\infty \\ &= Q, \end{aligned}$$

completing the proof of the theorem. ∎

LEMMA 25.2

Suppose P, Q, and R are projections in a von Neumann algebra $\mathcal{A}$. If $P \preceq Q$ and $Q \preceq R$, then $P \preceq R$.

PROOF Suppose $P \sim Q_1 \leq Q$ and $Q \sim R_1 \leq R$ with

$$P = U^*U,\ UU^* = Q_1 \leq Q, \qquad Q = V^*V,\ VV^* = R_1 \leq R,$$

where U and V are partial isometries in $\mathcal{A}$. Let $R_2 = VQ_1V^*$. Then R_2 is a projection with $R_2 \leq R_1$ $(\leq R)$, because $Q_1 \leq Q$ and $R_1 = VQV^*$. Let $W = VQ_1$. Then

$$W^*W = Q_1V^*VQ_1 = Q_1QQ_1 = Q_1$$

and

$$WW^* = VQ_1Q_1V^* = VQ_1V^* = R_2,$$

so that $P \sim Q_1 \sim R_2 \leq R$, or $P \preceq R$. ∎

THEOREM 25.3

For every von Neumann algebra $\mathcal{A}$ the relation $\preceq$ is a partial ordering on the set of equivalence classes of the projection lattice of $\mathcal{A}$.

PROOF This is a consequence of Lemmas 25.1 and 25.2. ∎

25.3 The Comparability Theorem

THEOREM 25.4

Let P and Q be projections in a von Neumann algebra $\mathcal{A}$. There exists a central projection Z such that $ZP \preceq ZQ$ and $(I - Z)P \succeq (I - Z)Q$.

PROOF Using Zorn's lemma we can find two maximal families $\{P_i\}_{i\in I}$ and $\{Q_i\}_{i\in I}$ of projections in $\mathcal{A}$ such that the P_i's are mutually orthogonal, the Q_i's are mutually orthogonal, $P_i \leq P$, $Q_i \leq Q$, and $P_i \sim Q_i$ for all $i \in I$. Let

$$P_0 = \sum_{i\in I} P_i,\ Q_0 = \sum_{i\in I} Q_i.$$

Then $P_0 \sim Q_0$ and by maximality there exist no nonzero projections $\widetilde{P} \leq P - P_0$ and $\widetilde{Q} \leq Q - Q_0$ such that $\widetilde{P} \sim \widetilde{Q}$. By Theorem 24.7 (recall that $Z(R)$ is the central support of R)

$$Z(P - P_0)Z(Q - Q_0) = 0.$$

Let $Z = Z(Q - Q_0)$. Then Z is a central projection with $Q - Q_0 \leq Z$ and $P - P_0 \leq I - Z$. It follows that

$$(P - P_0)Z = 0, \qquad (Q - Q_0)Z = Q - Q_0,$$

so that (note that $Q_0 \sim P_0$ and Z is central imply that $P_0 Z \sim Q_0 Z$)

$$PZ = P_0 Z \sim Q_0 Z \leq QZ$$

and (again $I - Z$ is central)

$$Q(I - Z) = Q_0(I - Z) \sim P_0(I - Z) \leq P(I - Z),$$

completing the proof of the theorem. ▮

COROLLARY 25.5

If $\mathcal{A}$ is a factor and P and Q are projections in $\mathcal{A}$, then one and only one of the following relations holds:

$$P \prec Q, \quad P \sim Q, \quad P \succ Q.$$

PROOF Let Z be a central projection guaranteed by the comparability theorem. Since $\mathcal{A}$ is a factor, either $Z = I$ or $Z = 0$. If $Z = I$, then $P \preceq Q$; if $Z = 0$, then $Q \preceq P$. So one and only one of $P \prec Q$, $P \sim Q$, $P \succ Q$ holds. ▮

Exercises

25.1 Show that $P_{2n} - P_{2n+1} \sim Q_{2n+1} - Q_{2n+2}$ and $Q_{2n} - Q_{2n+1} \sim P_{2n+1} - P_{2n+2}$.

25.2 Show that $P_\infty \sim Q_\infty$.

25.3 If $\{P_\alpha\}$ and $\{Q_\alpha\}$ are orthogonal families of projections in a von Neumann algebra such that $P_\alpha \preceq Q_\alpha$ for all α, show that $\oplus_\alpha P_\alpha \preceq \oplus_\alpha Q_\alpha$.

25.4 If P and Q are projections in a von Neumann algebra $\mathcal{A}$ and $P \preceq Q$, then $PR \preceq QR$ for every central projection R in $\mathcal{A}$.

26

Type Decomposition

26.1 Various Types of Projections

In this lecture we introduce three distinguished types of von Neumann algebras and show that every von Neumann algebra is the direct sum of algebras of these types.

DEFINITION 26.1 *Let $\mathcal{A}$ be a von Neumann algebra and P be a projection in $\mathcal{A}$.*

(a) *We say that P is finite if $P \sim Q \leq P$ implies that $P = Q$.*

(b) *We say that P is infinite if it is not finite.*

(c) *We say that P is abelian if $P\mathcal{A}P$ is abelian.*

(d) *We say that P is minimal if $P\mathcal{A}P = \mathbf{C}I$.*

It is clear that minimal projections are abelian. Also, if P is minimal, P majarizes no other nonzero projections. This is because $P \geq Q$ implies that $Q = PQP = \lambda I$, so that $Q = 0$ or $Q = P = I$.

Abelian projections are finite. In fact, if P is abelian and $P \sim Q \leq P$ with $V^*V = P$ and $VV^* = Q$, then for $U = PVP$ we have $U^*U = P$ and $UU^* = Q$. Since U and U^* are in $P\mathcal{A}P$ and $P\mathcal{A}P$ is abelian, we have $P = Q$.

The equivalence $\sim$ preserves finiteness. In fact, if P is finite, $V^*V = P$, $VV^* = Q$, (so that $P = V^*QV$) and $Q \sim Q_1 \leq Q$, then for $P_1 = V^*Q_1V \leq P$ and $U = Q_1V$ we have $U^*U = P_1$ and $UU^* = Q_1$, so that $P_1 \sim Q_1 \sim Q \sim P$, which implies that $P_1 = P$. Multiplying the equality $V^*Q_1V = V^*QV$ by V from the left and by V^* from the right, we obtain $Q_1 = Q$ and so Q is also finite.

If P is finite and $P_1 \leq P$, it is clear that P_1 is also finite.

The following proposition characterizes minimal projections in a von Neumann algebra.

PROPOSITION 26.1
Suppose $\mathcal{A}$ is a von Neumann algebra and P is a projection in $\mathcal{A}$. Then P is minimal if and only if $\mathcal{A}$ contains no proper subprojections of P.

PROOF Suppose P is minimal. Thus $P\mathcal{A}P = \mathbf{C}I$. If $Q \in \mathcal{A}$ is a subprojection of P, then $Q = PQP = aI$ for some $a \in \mathbf{C}$. The fact that Q is a projection forces $a = 0$ or 1. Thus Q is not a proper subprojection of P.

If $\mathcal{A}$ contains no proper subprojections of P, then $P\mathcal{A}P$ has no projections other than 0 and P (otherwise, such a projection would be a proper subprojection of P in $\mathcal{A}$). Since every von Neumann algebra is generated by its projections (see Theorem 20.3), we conclude that $P\mathcal{A}P = \mathbf{C}I$, or P is minimal. ∎

26.2 Centrally Orthogonal Projections

DEFINITION 26.2 *Suppose P and Q are two projections in a von Neumann algebra $\mathcal{A}$. We say that P and Q are centrally orthogonal if their central supports are orthogonal.*

Since every projection is majarized by its central support, centrally orthogonal projections are orthogonal.

LEMMA 26.2
The sum of centrally orthogonal abelian (or finite) projections $\{P_i\}_{i \in I}$ is again abelian (or finite).

PROOF For each $i \in I$ let $Z_i = Z(P_i)$ be the central support of P_i. Let $P = \sum_{i \in I} P_i$. Since $P_i \leq Z_i$ and the Z_i's are mutually orthogonal, we have $Z_i P_j = 0$ and $P_i = Z_i P$ for all $i \neq j$. For each T in $\mathcal{A}$

$$\begin{aligned} PTP &= \left(\sum_{i \in I} Z_i PT \right) P = \sum_{i \in I} Z_i PTZ_i P \\ &= \bigoplus_{i \in I} Z_i PTZ_i P = \bigoplus_{i \in I} P_i TP_i. \end{aligned}$$

If each $P_i \mathcal{A} P_i$ is abelian, then so is the direct sum

$$P\mathcal{A}P = \bigoplus_{i \in I} P_i \mathcal{A} P_i.$$

If each P_i is finite and $P \sim Q \leq P$, then $Z_i Q \sim Z_i P = P_i$ and

$$Z_i Q \leq Z_i P = P_i,$$

so that $Z_iQ = P_i$ for each i. This implies that

$$Q = QP = Q\sum_{i\in I} Z_iP_i = \sum_{i\in I} QZ_iP = \sum_{i\in I} P_iP = P,$$

so P is also finite. ∎

26.3 Type Decomposition

DEFINITION 26.3 *Suppose $\mathcal{A}$ is a von Neumann algebra.*

- ***(a)*** *If every nonzero central projection of $\mathcal{A}$ majarizes a nonzero abelian projection in $\mathcal{A}$, we say that $\mathcal{A}$ is of type I.*
- ***(b)*** *If $\mathcal{A}$ has no nonzero abelian projections but every nonzero central projection majarizes a nonzero finite projection, we say that $\mathcal{A}$ is of type II. A type II algebra is said to be of type II_1 if the identity operator I is finite. A type II algebra is said to be of type II_∞ if $\mathcal{A}$ has no nonzero finite central projections.*
- ***(c)*** *If $\mathcal{A}$ has no nonzero finite projections, we say that $\mathcal{A}$ is of type III.*

We can now prove the main result of this lecture.

THEOREM 26.3 THE TYPE DECOMPOSITION THEOREM
Every von Neumann algebra $\mathcal{A}$ acting on H is uniquely decomposable into the direct sum of von Neumann algebras of type I, type II_1, type II_∞, and type III.

PROOF Using Zorn's lemma we can obtain a maximal family $\{P_i\}$ of centrally orthogonal abelian projections in $\mathcal{A}$. Let $P = \sum P_i$. By Lemma 26.2 the projection P is again abelian. Let P_I be the central support of P and let $\mathcal{A}_I = P_I\mathcal{A}P_I$. Then $\mathcal{A}_I$ is a von Neumann algebra (acting on the Hilbert space $H_I = P_IH$) with identity P_I. Suppose Z is a nonzero central projection in $\mathcal{A}_I$ (majarized by P_I), then ZP is a nonzero abelian projection in $\mathcal{A}_I$ majarized by Z. This shows that $\mathcal{A}_I$ is of type I.

By the maximality of $\{P_i\}$, the algebra $(I - P_I)\mathcal{A}(I - P_I)$ has no nonzero abelian projections. Let $\{Q_j\}$ be a maximal family of centrally orthogonal finite projections in $(I - P_I)\mathcal{A}(I - P_I)$ and let $Q = \sum Q_j$. By Lemma 26.2 again Q is finite. Let P_{II} be the central support of Q. Since $Q \leq I - P_I$ and $I - P_I$ is central, we have $P_{II} \leq I - P_I$. Let $\mathcal{A}_{II} = P_{II}\mathcal{A}P_{II}$. Then $\mathcal{A}_{II}$ is a von Neumann algebra (acting on $H_{II} = P_{II}H$) with identity P_{II}. If Z is a nonzero central projection in $\mathcal{A}_{II}$ (majarized by P_{II}), then Z majarizes the nonzero finite projection ZQ, so that $\mathcal{A}_{II}$ is of type II.

By the maximality of $\{Q_j\}$, the projection $P_{III} = I - P_I - P_{II}$ does not majarize a nonzero finite projection; namely, the von Neumann algebra $\mathcal{A}_{III} = P_{III}\mathcal{A}P_{III}$ (acting on $H_{III} = P_{III}H$ with identity P_{III}) contains no nonzero finite projections, so that $\mathcal{A}_{III}$ is of type III. It is clear that

$$\mathcal{A} = \mathcal{A}_I \bigoplus \mathcal{A}_{II} \bigoplus \mathcal{A}_{III}.$$

Let $\{Z_k\}$ be a maximal family of orthogonal central finite projections in $\mathcal{A}_{II}$, let $P_{II_1} = \sum Z_k$, and let $P_{II_\infty} = P_{II} - P_{II_1}$. The von Neumann algebra $\mathcal{A}_{II_1} = P_{II_1}\mathcal{A}P_{II_1}$ is of type II_1 since its identity P_{II_1} is finite by Lemma 26.2. The maximality of $\{Z_k\}$ implies that $\mathcal{A}_{II_\infty} = P_{II_\infty}\mathcal{A}P_{II_\infty}$ has no nonzero central finite projections, so that $\mathcal{A}_{II_\infty}$ is of type II_∞. We now have the direct sum decomposition

$$\mathcal{A} = \mathcal{A}_I \bigoplus \mathcal{A}_{II_1} \bigoplus \mathcal{A}_{II_\infty} \bigoplus \mathcal{A}_{III}.$$

corresponding to

$$I = P_I \oplus P_{II_1} \oplus P_{II_\infty} \oplus P_{III}.$$

Suppose

$$I = P'_I \oplus P'_{II_1} \oplus P'_{II_\infty} \oplus P'_{III}$$

is another such decomposition. Then $P'_I(I - P_I) = 0$ because $I - P_I$ majarizes no nonzero abelian projections, $P'_I(I - P_I)$ is a central projection in $\mathcal{A}'_I$, and every nonzero central projection in $\mathcal{A}'_I$ must majarize a nonzero abelian projection. It follows that $P'_I \leq P_I$. By the same reasoning, $P_I \leq P'_I$, and so $P_I = P'_I$. It can be shown in a similar fashion that $P_{II_1} = P'_{II_1}$, $P_{II_\infty} = P'_{II_\infty}$, and $P_{III} = P'_{III}$, so that the type decomposition of $\mathcal{A}$ is unique. ∎

Note that in the decomposition above it may happen that some types are not present.

COROLLARY 26.4

A factor $\mathcal{A}$ is either of type I, type II_1, type II_∞, or type III.

PROOF Recall that the center of $\mathcal{A}$ is $\mathbf{C}I$. The only nonzero central projection of $\mathcal{A}$ is the identity. Thus it is not possible to decompose I into a direct sum of more than one nonzero central projections. ∎

Since the only nonzero central projection in a factor is the identity operator, we conclude that a factor is of type I if it has a nonzero abelian projection. A factor is of type II if it contains no nonzero abelian projection but contains nonzero finite projections; it is then of type II_1 or II_∞ according to whether the identity operator is finite or infinite. A factor is of type III if it contains no nonzero finite projections.

Exercises

26.1 If P is finite and $P_1 \leq P$, then P_1 is also finite.

26.2 If P is finite and $P_1 \preceq P$, then P_1 is also finite.

26.3 Show that $B(H)$ is a factor of type I.

26.4 Suppose $\mathcal{A}$ is a von Neumann algebra acting on H and P is a projection in the commutant $\mathcal{A}'$ of $\mathcal{A}$. Show that the commutant of $P\mathcal{A}P$ in $B(H_1)$ is $P\mathcal{A}'P$, where $H_1 = PH$.

26.5 Suppose $\mathcal{A}$ is a von Neumann algebra acting on H and P is a projection in $\mathcal{A}$. Show that the commutant of $P\mathcal{A}P$ in $B(H_1)$ is $P\mathcal{A}'P$, where $H_1 = PH$. Is the center of $P\mathcal{A}P$ in $B(H_1)$ equal to $P(\mathcal{A} \cap \mathcal{A}')P$?

26.6 Suppose $\mathcal{A}$ is a factor acting on H. Show that $\mathcal{A}$ is of type I if and only if $\mathcal{A}$ has a minimal projection, $\mathcal{A}$ is of type II if and only if it has no minimal projection but has nonzero finite projections, and $\mathcal{A}$ is of type III if and only if it contains no nonzero finite projections.

Bibliography

[1] Conway, J.B. *Subnormal Operators,* Pitman Research Notes in Math. **51**, Pitman, Boston-London, 1981.

[2] Douglas, R. *Banach Algebra Techniques in Operator Theory,* Academic Press, New York, 1972.

[3] Duren, P.L. *Theory of H^p Spaces,* Academic Press, New York, 1970.

[4] Garnett, J.B. *Bounded Analytic Functions,* Academic Press, New York, 1981.

[5] Kadison, R.V. and Ringrose, J.R. *Fundamentals of the Theory of Operator Algebras,* I, Academic Press, New York, 1983.

[6] Kadison, R.V. and Ringrose, J.R. *Fundamentals of the Theory of Operator Algebras,* II, Academic Press, New York, 1986.

[7] Murphy, G.J. *C^*-Algebras and Operator Theory,* Academic Press, London, 1990.

[8] Rickart, C.E. *General Theory of Banach Algebras,* Von Nostrand, New York, 1960.

[9] Rudin, W. *Functional Analysis* (2nd edition), McGraw Hill, Inc., New York, 1991.

[10] Rudin, W. *Fourier Analysis on Groups,* Interscience Publishers, New York, 1962.

[11] Takesaki, M. *Theory of Operator Algebras I,* Springer-Verlag, New York, 1979.

[12] Zygmund, A. *Trigonometric Series I and II* (2nd edition), Cambridge University Press, 1968.

Index

A page number followed by an asterisk * means that it is the page on which the indicated term is first defined or used. Usually this also means that the term (such as "Banach algebra") has so many occurrences in the book that it is inappropriate to list all the pages on which the term appears.